Lecture Notes in Artificial Intelligence   10183

Subseries of Lecture Notes in Computer Science

More information about this series at http://www.springer.com/series/1244

Friedhelm Schwenker · Stefan Scherer (Eds.)

Multimodal Pattern Recognition of Social Signals in Human-Computer-Interaction

4th IAPR TC 9 Workshop, MPRSS 2016
Cancun, Mexico, December 4, 2016
Revised Selected Papers

 Springer

Editors
Friedhelm Schwenker
Universität Ulm
Ulm
Germany

Stefan Scherer
Multimodal Communication
    and Computation
University of Southern California
Playa Vista, CA
USA

ISSN 0302-9743   ISSN 1611-3349   (electronic)
Lecture Notes in Artificial Intelligence
ISBN 978-3-319-59258-9   ISBN 978-3-319-59259-6   (eBook)
DOI 10.1007/978-3-319-59259-6

Library of Congress Control Number: 2017943009

LNCS Sublibrary: SL7 – Artificial Intelligence

Printed on acid-free paper

This Springer imprint is published by Springer Nature
The registered company is Springer International Publishing AG
The registered company address is: Gewerbestrasse 11, 6330 Cham, Switzerland

Preface

This book presents the proceedings of the 4th IAPR TC 9 Workshop on Pattern Recognition of Social Signals in Human-Computer-Interaction (MPRSS 2016). This workshop endeavored to bring recent research in pattern recognition and human-computer-interaction together, and succeeded to install a forum for ongoing discussions. In recent years, research in the field of intelligent human-computer-interaction has made considerable progress in methodology and application. However, building intelligent artificial companions capable of interacting with humans, in the same way humans interact with each other, remains a major challenge in this field. Pattern recognition and machine learning methodology play a major role in this pioneering research. MPRSS 2016 focused mainly on pattern recognition, machine learning, and information fusion methods with applications in social signal processing, including multimodal emotion recognition, user identification, and recognition of human activities. For the MPRSS 2016 workshop 13 out of 19 papers were selected for presentation at the workshop and for inclusion in this volume. MPRSS 2016 was held as a satellite workshop of the International Conference on Pattern Recognition (ICPR 2016) in Cancun, Mexico, on December 4, 2016.

This workshop would not have been possible without the help of many people and organizations. First of all, we are grateful to all the authors who submitted their contributions to the workshop. We thank the members of the Program Committee for performing the difficult task of selecting the best papers for this book, and we hope that readers of this volume may enjoy this selection of excellent papers and get inspired from its contributions. MPRSS 2016 was supported by the University of Ulm (Germany), the University of Southern California (USA), the Transregional Collaborative Research Center SFB/TRR 62 Companion-Technology for Cognitive Technical Systems at Ulm University, the International Association for Pattern Recognition (IAPR), and the new IAPR Technical Committee on Pattern Recognition in Human-Computer-Interaction (TC 9). Finally, we wish to express our gratitude to Springer for publishing our workshop proceedings in their LNCS/LNAI series.

March 2017
Friedhelm Schwenker
Stefan Scherer

Organization

Organizing Committee

Friedhelm Schwenker   Ulm University, Germany
Stefan Scherer   University of Southern California, USA

Program Committee

Shigeo Abe   Kobe University, Japan
Amir Atiya   Cairo University, Egypt
Mohamed Abdel Hady   Microsoft, USA
Markus Hagenbuchner   University of Wollongong, Australia
Hans A. Kestler   Ulm University, Germany
Nadia Mana   Fondazione Bruno Kessler, Italy
Günther Palm   Ulm University, Germany
Stefan Scherer   University of Southern California, USA
Friedhelm Schwenker   Ulm University, Germany
Eugene Semenkin   Siberian State Aerospace University, Russia

Sponsoring Institutions

International Association for Pattern Recognition (IAPR)
IAPR TC 9 on *Pattern Recognition in Human-Computer-Interaction*
Ulm University, Ulm, Germany
University of Southern California, USA
Transregional Collaborative Research Center SFB/TRR 62 *Companion-Technology for Cognitive Technical Systems*

Contents

Active Shape Model vs. Deep Learning for Facial Emotion Recognition in Security

Monica Bebawy[1], Suzan Anwar[2], and Mariofanna Milanova[2(✉)]

[1] Computer Science, Azusa Pacific University, Azusa, CA, USA
mbebawy12@apu.edu
[2] Computer Science, University of Arkansas at Little Rock, Little Rock, AR, USA
{sxanwar,mgmilanova}@ualr.edu

Abstract. As Facial Emotion Recognition is becoming more important everyday, A research experiment was conducted to find the best approach for Facial Emotion Recognition. Deep Learning (DL) and Active Shape Model (ASM) were tested. Researchers have worked with Facial Emotion Recognition in the past, with both Deep learning and Active Shape Model, with wanting to find out which approach is better for this kind of technology. Both methods were tested with two different datasets and our findings were consistent. Active shape Model was better when tested versus Deep Learning. However, Deep Learning was faster, and easier to implement, which means with better Deep Learning software, Deep Learning will be better in recognizing and classifying facial emotions. For this experiment Deep Learning showed accuracy for the CAFE dataset by 60% whereas Active Shape Model showed accuracy at 93%. Likewise with the JAFFE dataset; Deep Learning showed accuracy at 63% and Active Shape Model showed accuracy at 83%.

Keywords: Deep Learning · Active Shape Model · Facial emotion recognition · Neural network · Expression classification and Recognition

1 Introduction

Examination of facial expressions has a substantial significance in fields such as verbal, non-verbal expression and human-computer interface. Various approaches have been established in the Vision-based computerized expression recognition field lately. Fasel and Luettin and Pantic and Rothkrantz have studied these researches in detail in [6, 7]. As Paleari et al. proposes, it also conceivable progress multi-modal emotion recognition by the use of voice and visual data [8].

Murugappan et al. has accomplished to recognize emotions such as happiness, sadness, surprise and fear by use of time-frequency based methods of EEG data [9]. The emphasis of this study is to compare between existing software to test and see which approach is better for emotion recognition. Arı and Akarun, organized both skull movements and facial expressions by emerging face triangulation tracking based on high resolution, multi posture active object model and using the head trajectory information with Saklı Markov classification model

© Springer International Publishing AG 2017
F. Schwenker and S. Scherer (Eds.): MPRSS 2016, LNAI 10183, pp. 1–11, 2017.
DOI: 10.1007/978-3-319-59259-6_1

[10]. Akakin and Sankur have used sovereign mechanisms analysis results of the trajectory [11].

When evaluated with numerous classification methods, it is concluded that best result is obtained through use of 3D discrete cosine transform (DCT). Kuano et al. proposed a knowledge model for each emotion based upon variable concentration patterns and enabled face detection independent of the pose [12]. Sebe et al. has studied face expressions with Bayes nets, support vector machines and decision trees and presented the database they worked on to researchers [13]. Meanwhile, Littword et al. have recommended the systematic use of Adaboost, support vector machines and linear discriminant analysis methods [14].

Shan et al. used local binary patterns statistical model to classify the obtained attributes, using many artificial learning techniques, finally, Shan reported that support vector machines attained the best results [15]. Busso et al. used a commercial software for facial point tracking, which separates the human face in 5 areas of interest, calculated principal component analysis (PCA) coefficients of facial point locations for each region and classified this attribute vector obtained from these values with the nearest 3 neighbor [16].

Aggarwal and Shaohua Wan suggested an instinctive expression recognition method based on robust metric learning. They learnt a new metric space, the close data points have a higher likelihood of being in the same class, they also defined sensitivity and specificity to characterize the annotation steadfastness of each annotator [17]. Mao et al. anticipated a real time emotion recognition approach based on both 2D and 3D facial expression features extracting using Kinect sensors. They combined the features of animation units and feature point positions tracked by Kinect [18]. Suzan et al. implemented an Active Shape Model tracker, which tracks 116 facial landmarks. They used Support Vector Machine based classifier to recognize seven expressions. This technique is applied for the automated identification of the psychological state that exhibits a very strong correlation with the detected features [19]. Socher, Huval, Bhat, Manning and Ng Suggested that advances in 3D sensing technology made it easier to record color and depth images, which therefore, can improve object detection. This theory relies on convolutional neural networks and recursive neural networks (CNN and RNN) [20]. Le, Ngiam, Coates, Lahiri, Prochnow, and Ng proposed a new method to optimize Deep Learning. In their paper they show that more complex optimization methods can meaningfully simplify and accelerate the process of pertaining deep algorithms [21].

In this paper, a comparison between Active Shape Model (ASM) and Deep Learning (DL) for face emotion recognition is presented. In Detail, the following contribution is made.

First, a light on both Active Shape Model and Deep Learning is shed. Both approaches were tested and compared. By comparing Active Shape Model and Deep Learning gives the reader and future researchers the opportunity to see and decide which approach is better for their research and if they need to adjunct one or the other based on this research and its findings.

Secondly, Active Shape Model and Deep Learning were trained and tested with more than one database to make sure that our findings were accurate. The databases that were used included a variety of age, ethnic group, and gender to insure that both systems were well trained and will produce the best results.

Finally, facial emotion recognition can be used for security purposes. For example, using the Active Shape Model program that was provided, can calculate the percentage a person was anger, disgusted, and afraid; based on that a prediction of if that person is a threat to security or not.

The research begins with description of the dataset used and both Deep Learning as well as Active Shape Model proceeds with how they were trained. Then how both approaches were tested. The experiments done and the results obtained are given in Sect. 5. Section 6 summarizes the conclusions.

2   Background

Deep Learning (DL) and Active Shape Model (ASM) were compared to see which one is the better option for facial emotion recognition for security in real time. Both Deep Learning and Active Shape Model were tested with two datasets: The Japanese Female Facial Expression (JAFFE), and another dataset with children (CAFE). It was important to use more than one dataset, and to use a variety of age, gender, and ethnic group, because crime is not confined to a certain age, or one gender, nor one ethnic group.

2.1   Datasets

CAFE Dataset. The CAFE dataset was tested using LeNet with all 1192 pictures posed by different children. The findings were not very accurate.

JAFFE Dataset. First, the JAFFE dataset was tested using LeNet with all 213 pictures posed by ten different Japanese models. The findings were not very accurate.

2.2   Deep Learning

The software that was used for the Deep Learning software is the NVIDIA Deep Learning GPU Training System (DIGITS). This software provides three different neural networks for classification: LeNet, AlexNet, and GoogLeNet read more about them in [3–5]. To choose the best neural network to work with five pictures were taken of each dataset to test which neural network is the best for each dataset.

2.3   Active Shape Model

Active Shape Model (ASM) introduced by Cootes et al. is one of the most prevalent technique for detection and tracking of triangulation point. In this approach ASM is trained by introduction for tagged images. In order to identify the triangulation points in an image, first the location of face is detected with an overall face detector (such as Viola-Hones). The average face shape which is aligned according to position of the face constitutes the starting point of the search. Then the steps described below are repeated until the shape converges. (i) For each point, best matching position with the template is identified by using the gradient of image texture in the proximity of that point. (ii) The identified points are projected from their point locations in training set to the shape eigenvalues which is obtained by Principal Component Analysis (PCA). Whereas the individual template matchers in the first step may diverge from their sound positions and the shape obtained may not look like a face, the holistic approach used in the second step strengthens the independent weak models by constraining them and associating them with the shapes in the training set.

3   Training

3.1   Deep Learning

The training of the Deep Learning software started with two datasets: JAFFE and CAFE.

CAFE Dataset. First the CAFE dataset was used to test the Deep Learning software. To choose between the three neural network five random pictures were selected from each emotion and tested them with the neural networks available. LeNet was better than AlexNet by 40% which was better than GoogLeNet by 65%. Here is a figure of the training graph for the LeNet for the CAFE dataset.

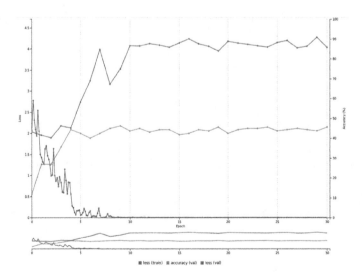

Fig. 1. This is the training graph for the Cafe Dataset we see that the blue is the loss, the orange is the accuracy, and the green is the loss in the training sample. (Color figure online)

In Fig. 1: the loss(Blue) is decreasing as the neural network is given more pictures to train with, which is what makes this neural network the best one for our dataset.

JAFFE Dataset. Then, the JAFEE dataset was used to test the Deep Learning software. The same test was run to choose the neural network as the CAFE dataset. LeNet was better then GoogLeNet by 43% which was better than AlexNet by 12%. Here is the figure of the training graph foe the LeNet for the JAFFE dataset.

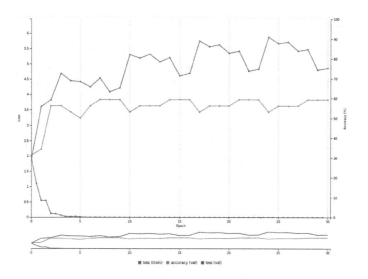

Fig. 2. This is the training graph for the JAFFE Dataset. The blue is the loss, the orange is the accuracy, and the green is the loss in the training sample. (Color figure online)

In Fig. 2: Similar results are shown to the CAFE dataset, the blue (loss) is decreasing and it stays around *zero*. Again, this is what makes this neural network the best one to work with the JAFFE dataset.

3.2 Active Shape Model

When ASM need to be extended for training of a new subject, photographs of the individual are taken and triangulation points of this photograph are marked up automatically by matching with general model. After this marked points are fine-tuned with Pinotator [22] an ASM specific for the person is generated. Then classifier can be easily trained for new person and environment, enabling use of the system for new people. Since facial expressions differ for each person and the environment (such as lighting), here a mechanism which can configured according to a specific person and specific environment has been targeted. In this mechanism subjects start with a neutral expression and wait for 2 s. In this process an average of derived attributes is calculated and in frames following that, in order to enable system to act independent of environmental variables, attributes are normalized by division to their average. Each individual who will use the system repeats each expression T times, recording a total of NT samples (N = number of expression classes).

4   Testing

4.1   Deep Learning

The NVIDIA DIGITS provided the research with very effective information about Deep Learning and how it is used [1]. The NVIDIA DIGITS was used to test facial emotion recognition with multiple datasets, The Japaneses Female Facial Expression Database (JAFFE) was the first dataset that was used. Every dataset was trained with three different networks: LeNet, AlexNet, and GoogLeNet.

Every dataset reacted differently when tested with the different networks; however, in our case both datasets worked best with LeNet.

CAFE Dataset. Here is a table showing the findings and the accuracy percentage.

Table 1. The results for the CAFE dataset tested with Deep Learning using LeNetwork

Emotion	Total pictures	Right classification	Percentage
Anger	205	95	46%
Disgust	191	139	73%
Fear	140	66	47%
Happiness	215	147	68%
Neutral	230	188	82%
Surprise	103	62	60%
Sadness	108	50	46%

In Table 1 it is shown that the accuracy is not very good. However, it was very hard to test the CAFE dataset. Since it is using children as the models for the pictures it is very hard to get the models to understand and demonstrate the emotion without having any confusion. In the Active Shape Model however, the accuracy is a lot higher as it will be shown in the next section.

JAFFE Dataset. Here is a table showing the findings and the accuracy percentage.

In Table 2 it is shown that the accuracy is a lot better, since the models are all adults and therefore can clearly show the labeled emotion.

Table 2. The results for the JAFFE dataset tested with Deep Learning using LeNetwork

Emotion	Total pictures	Right classification	Percentage
Anger	30	17	57%
Disgust	29	17	59%
Fear	32	25	78%
Happiness	31	25	81%
Neutral	30	14	47%
Surprise	30	19	63%
Sadness	31	17	55%

4.2   Active Shape Model

CAFE Dataset. First, the Active Shape Model computer program was tested with the CAFE dataset to compare it with the Deep Learning results.

Table 3. The results for the CAFE dataset tested with Active Shape Model

Emotion	Total pictures	Right classification	Percentage
Anger	205	188	92%
Disgust	191	179	94%
Fear	140	132	94%
Happiness	215	200	93%
Neutral	230	225	98%
Surprise	103	94	91%
Sadness	108	98	91%

It is clear that Active Shape Model shows a much better accuracy with the CAFE dataset than Deep Learning with LeNet the total percentage for the CAFE dataset using ASM is 93% which is very accurate given all the challenges (Table 3).

During the test phase, the subject once more starts with a neutral expression, prepares the system to initial state to avoid disturbances of the environment. In each following frame, attribute vector for the face is calculated. The average distance value of each element in the attribute vector to each class is calculated which is di, i = 1,....N vectors. The distance values depict differences, whereas Si = e − di values depict similarities and used as similarity metrics. Finally Si vector is normalized as the sum of their elements will equal to 1. This values can be thought as a probability for each class.

5   Findings

After testing both Deep Learning and Active Shape Model, Active Shape model was found to be indeed better in facial emotion recognition in the CAFE dataset Deep Learning was accurate by 60% but Active Shape Model was accurate by 93%.

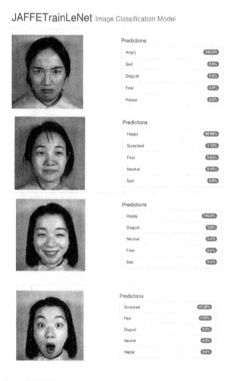

Fig. 3. Results for the JAFFE dataset when tested with Deep Learning DIGITS system.

In Fig. 3: it is shown how the DIGITS system works after training the system, this is the result that was obtained when testing it with these four pictures. It was pretty accurate with these pictures, however, for the second picture the picture was not labeled as happy, nevertheless, the model does show some happy features.

In Fig. 4: it shows when the CAFE dataset was tested with the Active Shape Model. Most of these pictures are classified correctly.

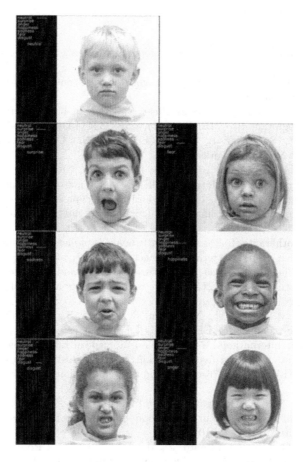

Fig. 4. Results for the CAFE dataset when tested with Active Shape Model system.

6   Conclusion

Facial Emotion Recognition is very important and because of that an experiment was conducted to compare between Active Shape Model and Deep Learning. This experiment used NVIDIA DIGITS software and a computer program by Ms. Anwar. Based on the previous sections it was concluded that Active Shape Model is in fact better than Deep Learning given the specific computer programs that were used. However, this experiment could be conducted with different software to produce different results. Since many variables were used, the same experiment with more controlled variables might provide different results.

Acknowledgment. This project is supported by the National Science Foundation under the award CNS1359323 This project is funded by the National Science Foundation for Undergraduate students. (https://sites.google.com/a/ualr.edu/cs-reu-site-ualr/).

References

1. NVIDIA DIGITS (2015). https://developer.nvidia.com/digits, Accessed 08 Jul 2016
2. CyberSAFE@UALR (n.d.). https://sites.google.com/a/ualr.edu/cs-reu-site-ualr/, Accessed 08 Jul 2016
3. Lecun, Y., Bottou, L., Bengio, Y., Haffner, P.: Gradient-based learning applied to document recognition. Proc. IEEE **86**(11), 2278–2324 (1998). doi:10.1109/5.726791
4. Krizhevsky, A., Sutskever, I., Hinton, G.E.: ImageNet Classification with Deep Convolutional Neural Networks (n.d.)
5. Szegedy, C., Liu, W., Jia, Y., Sermanet, P., Reed, S., Anguelov, D., Rabinovich, A.: Going deeper with convolutions. In: IEEE Conference on Computer Vision and Pattern Recognition (CVPR) ((2015)). doi:10.1109/cvpr.2015.7298594
6. Fasel, B., Luettin, J.: Automatic facial expression analysis: a survey. Pattern Recogn. **36**, 259–275 (2003)
7. Pantic, M., Rothkrantz, L.J.M.: Automatic analysis of facial expressions: the state of the art. IEEE Trans. Pattern Anal. Mach. Intell. **22**, 1424–1445 (2000)
8. Paleari, M., Chellali, R., Huet, B.: Features for multimodal emotion recognition: an extensive study. In: IEEE Conference on Cybernetics and Intelligent Systems, pp. 90-95 (2010)
9. Murugappan, M., Rizon, M., Nagarajan, R., Yaacob, S., Hazry, D., Zunaidi, I.: Time-frequency analysis of EEG signals for human emotion detection. In: 4th Kuala Lumpur International Conference on Biomedical Engineering (2008)
10. Arı, İ., Akarun, L.: Facial feature tracking and expression recognition for sign language, In: IEEE, Signal Processing and Communications Applications, Antalya (2009)
11. Akakın, H.Ç., Sankur, B.: Spatiotemporal features for effective facial expression recognition. In: IEEE 11th European Conference on Computer Vision, Workshop on Sign Gesture Activity (2010)
12. Kumano, S., Otsuka, K., Yamato, J., Maeda, E., Sato, Y.: Pose-invariant facial expression recognition using variable-intensity templates. Int. J. Comput. Vis. **83**, 178–194 (2008)
13. Sebe, N., Lew, M.S., Sun, Y., Cohen, I., Gevers, T., Huang, T.S.: Authentic facial expression analysis. Image Vis. Comput. **25**, 1856–1863 (2007)
14. Littlewort, G., Bartlett, M.S., Fasel, I., Susskind, J., Movellan, J.: Dynamics of facial expression extracted automatically from video. Image Vis. Comput. **24**, 615–625 (2006)
15. Shan, C., Gong, S., McOwan, P.W.: Facial expression recognition based on local binary patterns: a comprehensive study. Image Vis. Comput. **27**, 803–816 (2009)
16. Busso, C., Deng, Z., Yildirim, S., Bulut, M., Lee, C.M., Kazemzadeh, A., Lee, S., Neumann, U., Narayanan, S.: Analysis of emotion recognition using facial expressions, speech and multimodal information. In: Proceedings of the 6th International Conference on Multimodal Interfaces - ICMI 2004, New York (2004)
17. Shaohua, W., Aggarwal, J.K.: Spotaneous facial expression recognition: a robust metric learning approach, Computer Vision Research Center, The University of Texas at Austin, Austin, TX 78712–1084. US, Pattern Recognition 47 (2014)
18. Mao, Q., Xinyu, P., Zhan, Y., Xiangiun, S.: Usinng Kinect for real time emotion recognition via facial expression. Front. Inf. Technol. Electron. Eng. **16**(4), 272–282 (2015)

19. Anwar, S., Milanova, M., Bigazzi, A., Bocchi, L., Guazzini, A.: Real Time Intention Recognition IEEE IECON 2016, Florence, 24–28 October 2016
20. Socher, R., Huval, B., Bhat, B., Manning, C.D., Ng, A.Y.: Convolutional-Recursive Deep Learning for 3D Object Classification, pp. 1–9 (n.d.)
21. Le, Q.V., Ngiam, J., Coates, A., Lahiri, A., Prochnow, B., Ng, A.Y.: On Optimization Methods for Deep Learning (n.d.)
22. Arı, İ., Açıköz, Y.: Fast image annotation with Pinotator. In: IEEE 19th Signal Processing and Communications Applications Conference (2011)

Bimodal Recognition of Cognitive Load Based on Speech and Physiological Changes

Dennis Held, Sascha Meudt$^{(\boxtimes)}$, and Friedhelm Schwenker

Institute for Neural Information Processing, Ulm University, 89069 Ulm, Germany
{dennis.held,sascha.meudt,friedhelm.schwenker}@uni-ulm.de

Abstract. An essential component of the interaction between humans is the reaction through their emotional intelligence to emotional states of the counterpart and respond appropriately. This kind of action results in a successful interpersonal communication. The first step to achieve this goal within HCI is the identification of these emotional states.

This paper deals with the development of procedures and an automated classification system for recognition of mental overload and mental underload utilizing speech an physiological signals. Mental load states are induced through easy and tedious tasks for mental underload and complex and hard tasks for mental overload. It will be shown, how to select suitable features, build uni modal classifiers which then are combined to a bimodal mental load estimation by the use of early and late fusion. Additionally the impact of speech artifacts on physiological data is investigated.

1 Introduction

The interaction between humans and machines is ubiquitous in today's world. Whether at the station, while solving a web card or when using a mobile phone, it has become the aim of making this interaction between man and machine to the user so intuitive, effective and pleasant as possible. While this is desirable in most cases, but not always reach satisfactory levels. A key factor that affects the human interaction significantly, has not yet been taken into account - the emotional intelligence [6]. By the term emotional intelligence one understands the ability of the regulation, the use of knowledge and the expression of emotions in interactions with others or with themselves. A system that is centered on the satisfaction of persons needs should offer these expertise to react appropriately to the emotional state [10].

Recognition of affective states on multiple modalities, such as facial expressions, speech, gestures, and to a lesser extent on physiological changes. The recognition of these states is possible because by state changes the expressions adjust accordingly such as the pitch of the voice when is excited.

The consideration in this work are speech and physiological signals. The speech is in terms of research a dominant modality, as for the people it is representing the most natural way of communicating and a very quick indicator for an emotional state of the counterpart. A closer look at speech turns out that not

© Springer International Publishing AG 2017
F. Schwenker and S. Scherer (Eds.): MPRSS 2016, LNAI 10183, pp. 12–23, 2017.
DOI: 10.1007/978-3-319-59259-6_2

only the content of what is said is relevant, but also the way on how its said and which emotions are placed in an utterance. The recognition of the content by machines, with a high detection rate of about 90%, is already realized, however, the rate of recognition of speech emotions is only about 60% [11]. Physiological signals like electrocardiography (ECG), electromyography (EMG) and electro-dermal activity (EDA) provide a relatively new and growing area of research in terms of affective state recognition compared to audiovisual emotion recognition. The advantages of physiological signals are that the regulation of the values such as heart rate or the activity of the sweat glands cannot be simply consciously influenced, such as speech and gestures. This leads to an almost undistorted image on the emotional state of the person.

In the past there had been a lot of research on detecting emotional states utilizing single modalities, like detecting affect from speech with detection rates from 50% up to 90% [7,8]. For physiological signals there are many studies on the impact and the detection of mental load [2,3]. For bimodal approach with speech and physiological signals J. Kim shows an improvement in recognition rate compared to unimodal classification [4]. Most of the studies are based on acted or even strong expressive emotional datasets on basic emotions for example defined by Eckman [1]. This methods can not be easily transferred to typical weak expressive behaviours in HCI which in addition typically does not refer to basic emotions.

The utilized dataset contains data from natural behaving, none acting, users interacting with a HCI system which is able to induce different mental load levels. The investigated affective states are mental overload and underload that both have a negative impact on the performance of a person. Firstly, even simple tasks work in a period of excessive demand as relatively difficult. On the other tasks are perceived as boring during a mental underload, which leads to a lack of concentration and thus to a decrease in performance. Based on the dataset this paper develops an classification system for these affective states with the use of ensemble classifiers. The interests are the unimodal classification based on speech and physiological signals, the fusion of these to a bimodal classification result and the investigation of the influence of speech artifacts on the classification of physiological signals.

2   Experimental Setting

The dataset is based on an experiment done within the Transregional Collaborative Research Centre SFB/TRR 62 "Companion-Technology for Cognitive Technical Systems".

Participants were asked to play a series of games based on the Interaction paradigm of Schüssel et al. [12]. The task of each game sequence was to identify the singleton element, i.e. the one item that is unique in shape and color (number 36 and 2 in Fig. 1). The difficulty was set by adjusting the number of shapes shown and the maximum time to answer. If the given answer was incorrect, they received no reward for that particular round. After an introduction each participant completed five game sequences of decreasing difficulty. The first sequence

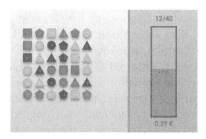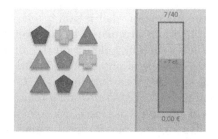

Fig. 1. Screen shot of the difficult level (left) with target element 36 and the easiest one (right) with target element 2. (Color figure online)

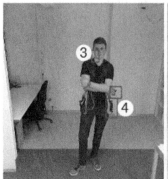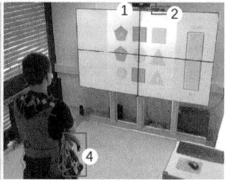

Fig. 2. Overview of the setting with sensors: (1) MS Kinect 2, (2) frontal webcam, (3) wireless headset, (4) GTec g.MOBIlab+ biophysical sensor with sensors attached to the users body.

was designed to induce overload (6×6 board, 6 s to answer, see Fig. 1 left), the second was 5×5 with 10 s, the third was set to 3×3 with 100 s, sequence four again was 3×3 mode with 100 s time (underload). As the sequences 1 and 4 are explicitly designed to cause mental overload and underload, we focused on those two.

After each sequence played, the participants answered a Self Assessment Scale questionnaire (SAM). The aim of those questions was to determine valence, arousal and dominance experienced in the particular sequence. A total of 60 participants were recorded. Of those were 30 male and 30 female. Their age spanned from 17 to 27 (mean 21.97, $\sigma^2 \approx 2.6$).

During the experiment, participants were monitored by several sensors providing multimodal synchronous data. See Fig. 2. The sensory system contains two webcams (Logitech C9100), one in front locking towards the users face and one from the rear providing an overview of the scene, a wireless headset, a Mirosoft Kinect v2 camera in front recording rgb, infrared, depth, skeleton/postural and audio information and finally an GTec g.MOBIlab+ biophysical sensor recording ECG, EDA, EMG (trapezius muscle), Respiration and Temperature. In this work we focus on the audio and physiological data.

3   Unimodal Recognition of Cognitive Load

At first, two unimodal classifiers for speech and two for every channel of the physiological signals were trained. This results in a total of twelve ensemble classifiers, where six are specialized on mental overload (OL) and the other six are trained for recognising the amount of mental underload (UL). These unimodal classifiers are evaluated using individual classification (SELF) method with a 10×10-fold cross validation and leave one subject out (LOSO) method with 10-fold cross validation.

In Chap. 4 fusion approaches for the bimodal classification are shown and the influence of speech on the classification of physiological signals.

3.1   Speech

For cognitive load recognition from speech, random forest ensembles were used to classify. Only the utterances contains useful parts without speaking breaks (silence) are dictated by the experimental settings, this reduces the average length of recorded overload sequence from 362 to 22 s. and for the underload sequence from 324 to 18 s. The average amount of utterances within the overload sequence is 45 and for the underload sequence it is 43.

The extracted features of the audio data is subdivided into the feature extraction methods:

- Linear predictive filter coefficients (LPC)
- Mel-frequency cepstral coefficients (MFCC)
- Relative spectral perceptual linear prediction (Rasta-PLP)
- Modulation spectrum (ModSpec)

The window size for each frame ranges between the different feature extraction methods from 40 to 200 ms. The window shift is 20 ms for every feature instance. Overall for every frame 57 features are extracted consisting of 8 LPC, 20 MFCC, 21 Rasta-PLP and 8 ModSpec.

The evaluation with the SELF and LOSO method results in Table 1 show the average classification accuracy based on different features and the early fusion of these features. Classification based on utterances, containing an aggregation of severe frame level decisions, for SELF and LOSO method shows better average classification accuracies than based on single frame level. The highest result with respect to accuracy are achieved through early fusion of the features and the classification of utterances for the SELF and LOSO method.

3.2   Physiology

In order to recognize mental overload on the basis of physiology, a random forest classifier for each of the following physiological signals was trained: Electrocardiography, trapezius trapezius Electromyography, Electrodermal activity, respiration and body temperature. For the physiological signals 58 statistical features

Table 1. Evaluation results of all speech classifiers using SELF and LOSO method for every used extraction method. The results are subdivided into classification results based on frame and utterances level.

Analysis	SELF		LOSO	
	Frame	Utterance	Frame	Utterance
LPC	66.52%	77.75%	61.22%	68.93%
MFCC	76.27%	83.80%	68.54%	75.53%
Rasta-PLP	74.49%	82.75%	68.70%	74.40%
ModSpec	70.20%	78.82%	63.90%	71.29%
Audio-Fusion	81.67%	86.58%	72.72%	77.50%

were extracted, which results in 290 features over all. The features are extracted on 5 s data chunks with an overlap of 4.9 s. The preprocessing for the physiological signals is subdivided into two groups, one for ECG and one for the rest of the physiological signals.

For the ECG signal the preprocessing consists of linear detrending and the normalization of the signals with the mean R-peaks, because the sensors are not exactly at the same position at each participant. This creates a different mean peak value for every participant which has no information about the mental load of a participant. For EMG, EDA, respiration and temperature the preprocessing consists of the use of a butterworth filter with low and high cutoff frequencies of 10 Hz and 125 Hz and the order of four.

The features extracted from the ECG signal are based on 25 features from wavelets [15] and 33 statistical features from the PQRST complex [13]. Features extracted from EMG, EDA, respiration and temperature are based on statistical and mathematical features in time and fequency domain [5,9,14].

All four classifiers and an additional fusion of these classifiers were evaluated using the SELF and LOSO method. Table 2 shows the average classification accuracies over all 60 participants. The weights for the fusion which are used to fuse the outputs together are calculated through a Moore-Penrose pseudoinverse. The SELF classification results for ECG are promising and achieve an average classification accuracy of 91.52%. The EMG channels shows that there are three different activity patterns for the trapezius muscle:

– More activity in the overload sequence (e.g. in Fig. 3 left)
– less activity in the overload sequence (e.g. in Fig. 3 right)
– almost the same activity in the overload and underload sequence

The results show that the difference between mental overload and underload and therefore the classification can be highly accomplished for SELF, such that the person has a characteristic behaviour for these affective states. Figure 3 shows that for both participants on their own the development of the MAV feature is easily distinguishable between OL and UL sequence. But for the LOSO

Table 2. Evaluation results of all physiological classifiers using SELF and LOSO method for every used extraction method.

Analysis	SELF	LOSO
ECG	91.52%	66.53%
EMG	74.53%	51.03%
EDA	64.68%	51.51%
Respiration	64.68%	51.50%
Temperature	60.89%	54.23%
Fusion	91.91%	65.61%

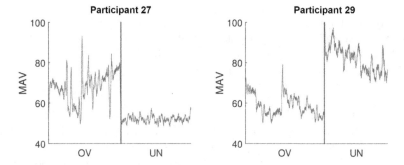

Fig. 3. Mean absolute value (MAV) feature from the EMG signal of two participants. Shows the course of MAV feature in the overload (OL) and underload (UL) sequence.

classification Fig. 3 and the results in Table 2 show that there is not an general feature expression in respect to the physiological signals.

4 Bimodal Recognition of Mental Overload and Underload

In order to achieve bimodal classification results, two different fusion approaches were used to aggregate the unimodal results. The input consists of two classification results from the speech classifiers for (OL and UL) and two classification results from the physiological classifiers. The output of these classifiers is continuous between 0 and 1, e.g. where 1 for an OL classifier corresponds to that every weak learner classified it as mental overload.

For the bimodal evaluation the data was taken from both modalities, if a participant speaks. If there is no classification from speech the classification result is just based on the physiological classification. The classification structure which is used produces for every utterance for both, the physiological and the audio signal, one classification result.

4.1   Moore-Penrose Pseudoinverse

The pseudoinverse is used to create weights for the OL and UL classifier. Two calculated weights are needed. One for the OL where the extend of mental overload is classified and one for the UL where the extend of mental underload is classified. For the OL classifier a mental overload frame should be classified as 1 and an underload frame should be classified as 0 but for the UL classifier it should work the other way around such that for an underload frame it should be 1.

For the calculation of the weights e.g. for the OL classifier follows these steps:

1. Train the ensembles
2. Take a train set for the fusion with N frames which are not used to train the ensemble system and which have no intersection with the testset
3. Classify the train set for the fusion thought the trained ensemble system, results in two unimodal classification results
4. X ($N \times 2$) for OL (two stands for both modalities)
5. Calculate the pseudoinverse $X^+ = (X^t X)^{-1} X^t$
6. Calculate the weights $w = X^+ Y$

The same steps should be done for the UL part of the classification system. After theses steps the weights for the bimodal fusion layer are set. The calculated weights strongly depend on the data presented to train the fusion layer. For the SELF classification there are 284 frames and for the LOSO there about 16450 frames to train this layer. This corresponds to about 11% of the data used to train the ensembles.

Numerical Evaluation. Table 3 shows the classification accuracy, for SELF classification it is 94.07%. The weights for the input channels through the pseudoinverse for the SELF classification have a $\sigma^2 = 0.38$ and a mean weight for speech of 0.23 and for physiological of 0.77. The difference of the weights for the modalities is based on the greater variance of the physiological classification between participants.

For the LOSO classification the weight for speech is 0.76 and for physiological it is 0.24. The variance of the weights for LOSO classification is far smaller because only about 1.7% of the data within the training set changes for the classification for a new participant. The overall average accuracy for the LOSO classification is 76.31%. The difference for the weights between the OL and the UL classifier are -0.01 for the speech and $+0.01$ for the physiological inputs.

4.2   Modified Max-Voting

The second fusion method used is a modified max-voting (MMV). This method does not have any learning phase and can be used immediately to fuse the unimodal classification results. The input consists, as described, of two values for OL and two for UL. For the OL and the UL classification results from the different modalities are averaged. This results in two confidences, one for OL and one

Table 3. Bimodal classification accuracy with the Moore-Penrose pseudoinverse method for SELF and LOSO classification.

Analysis	SELF	LOSO
Pseudoinverse	94.07%	76.31%
OL	95.59%	75.80%
UL	92.55%	76.82%

for UL for each modality. The last step calculates a MMV on these OL and UL classifiers and the bimodal classification result is based on the higher confidence of OL or UL. The bimodal classification confidence is calculated through the difference of confidence of the OL and UL within the step before.

For this fusion method, weights for audio and physiology classification results are the same. It just differentiates how confident the bimodal OL and UL classifiers are. It takes advantage of the classification result for one modality if it has an higher confidence than the confidence of the classifier for the other modality. The more secure a classifier is that it is a specific affective state the less important the other modality is.

Numerical Evaluation. Table 4 shows that SELF classification has an average accuracy of 96.41%. The difference between the recognition of mental overload to underload is 3.04%. For 83% of participants the accuracy improves from the unimodal classification results to the bimodal results from 1% up to 29%. This effect is particularly pronounced for a participant which has the lowest classification rate of 67.7% for the physiological signals. His rate increases through the bimodal classification to 95.4%. Another effect is the decrease of variance for the classification results for the different participants: audio ($\sigma^2 = 2.9$), physiological ($\sigma^2 = 5.6$) and bimodal fusion through MMV ($\sigma^2 = 1.6$).

LOSO classification shows an average accuracy of 72.55% but a greater difference between the recognition of mental overload to underload. With a difference of 21.41% between OL and UL it is far more unbalanced in respect to the fusion approach with pseudoinverse.

Table 4. Bimodal classification accuracy with the modified max-voting method for SELF and LOSO classification. The accuracies for the mental overload and underload are shown.

Analysis	SELF	LOSO
MMV	96.41%	72.55%
OL	97.93%	82.56%
UL	94.89%	61,15%

Table 5. Evaluation results of all physiological classifiers and the comparison between the classification of physiological signals while the participants speak and does not speak using SELF and LOSO method.

Analysis	SELF		LOSO	
	Speaking	No speaking	Speaking	No speaking
ECG	91.08%	92.72%	65.70%	67.85%
EMG	74.54%	75.19%	49.55%	53.62%
EDA	64.18%	66.08%	49.95%	54.15%
Respiration	57.78%	60.27%	49.72%	54.73%
Temperature	60.47%	62.27%	53.47%	55.65%
Fusion	91.73%	93.00%	65.82%	68.42%

4.3   Influence of Speech to Physiological Classification

The investigation deals with the influence of speech in the classification of physiological signals. The aim is to discover the impact on different physiological channels like EMG and conclude how serious these impacts are.

The change in physiological signals through speech is based on two factors. The first is the direct influence for example through the vibration of the vocal cords or the change of air volume within the lungs and therefore influence on the physiological signal values like for the respiration signal a higher measured respiration rate while speaking. The second is the real influence if there is an direct influence through speaking on the physiological signals e.g. a person speaks if there is significant influence to change the physiological signals and therefore are not based on the emotion to recognize rather than on speaking itself. For example if speaking has a negative influence on the recognition of emotion through physiological signals the reason for this could be that if somebody speaks there is an physiological signature for this within the physiological signals and therefore in all sequence similar. The reason for an influence within the signal is hard to assign based on these two influences therefore it is investigated to what extend the values are changing while speaking and if there is an positive, negative or no influence on the classification.

The training and evaluation is done with extracted frames from the physiological signals where the participant did not speak in any of the 5 s length of these frames against the others where the physiological signals are influenced by speaking. This is done for all physiological channels as well as for the unimodal fusion of these channels.

Numerical Evaluation. Table 5 shows the classification result for the physiological signals with and without any speaking involved. For the SELF classification the average classification accuracy improves for ECG by 1.64%, EMG by 0.65%, EDA by 1.90%, respiration by 2.49%, temperature by 1.80% and for fusion by 1.27%. 42 participants improve their SELF classification through the

Table 6. T-Test between speaking and not speaking for SELF and LOSO classification of the physiological signals.

T-Test	SELF	LOSO
	Speaking/no speaking	Speaking/no speaking
ECG	0.004	0.002
EMG	0.465	$4.4\,e^{-5}$
EDA	0.041	$2.2\,e^{-4}$
Respiration	0.007	$2.3\,e^{-10}$
Temperature	0.009	$4.0\,e^{-4}$
Fusion	0.003	$4.9\,e^{-9}$

fusion up to 8.18%. The improvements are based primarily on the change of classification accuracy of mental underload (+1.52%).

For the LOSO classification the average classification accuracy improves for all channels: ECG by 2.15%, EMG by 2.59%, EDA by 4.07%, respiration by 4.20%, temperature by 2.18% and for the fusion by 2.60%. 47 participants improve their LOSO classification through fusion up to 12.84%. The increase of accuracy is based on the improvement of the recognition for both mental overload and underload. For both SELF and LOSO classification the improvements are the same for male and female.

To investigate if there is a statistical significant difference between the classification of all physiological signals if the participant speaks and did not speak a T-Test is used. The results are shown in Table 6. The results support that there is an difference between speaking and not speaking. For example at an $\alpha = 0.05$, all null hypothesis are rejected except the EMG for SELF classification. The highest support for an difference show the respiration channel with LOSO classification.

4.4   Discussion

The investigation of the influence of speech artifacts while the classification of physiological signals shows that the classification accuracy improves both for SELF and LOSO classification if parts containing speech are removed from the physiological data pool. This could be based on the two influences described in Subsect. 4.3. The first could be due to movement of the subject which influences the physiological signal like noise. The second could have more impact on the change in classification results based on my observation. The reason for this is that the participants behave during and shortly after speaking with respect to their posture and activity differently between speaking and not speaking. This change in behaviour while speaking, however is noted in both sequences almost equally, which means that a similar behaviour in the OL and UL sequence is shown and therefore the meaningfulness of the physiological signals while speaking are not so depend on the emotional state as the participant does not speak. The reason might be clear especially towards the end of the utterance that the participants are waiting for the response of the system, so if their answer was accepted and whether the answer was correct or incorrect.

For SELF classification shows that the MMV is 2.34% better than the fusion with pseudoinverse. A reason for this could be that the weights for pseudoinverse method are dependent on the trainings set for this fusion layer and the training set could show another kind of behaviour than the rest of the data. This could lead into questionable weights for both modalities. For the LOSO classification the pseudoinverse is 3.76% better than the fusion with MMV. The classification of the underload sequence is ≈15% better with the pseudoinverse. The reason for this could be the amount of data which could be used to train this method. Because about 8.3% of the data from 59 out of 60 participants could be used to train the bimodal fusion layer. The probability to use data to train the bimodal fusion layer that does not represent the remaining data to train the ensembles is shrinking compared to the SELF classification.

The LOSO accuracies for the speech data achieve a higher accuracy than the bimodal classification and the reason for this could be the high deviation of classification results from the physiological data. The LOSO classification accuracies for the physiological data reaches from 35.72% up to 89.62%. The reason for this could be the different behaviour patterns for the participants. The classification of participants which have the behaviour pattern of no different movement within the overload and underload sequence improves by 1.91%. The amount of participants which have more activity within the overload sequence are approximately double as frequent as participants which have more activation within the underload sequence. This results in the lower accuracy for the classification of participants with lower acitivity within the overload sequence for the physiological signals.

The bimodal classification results show that for the SELF classification average classification accuracy of 96.41% could be achieved and for the LOSO classification 76.31%.

Acknowledgments. The authors of this paper are partially funded by the Transregional Collaborative Research Centre SFB/TRR 62 "Companion-Technology for Cognitive Technical Systems" funded by the German Research Foundation (DFG).

References

1. Ekman, P.: An argument for basic emotions. Cogn. Emot. **6**(3–4), 169–200 (1992)
2. Gevins, A., Smith, M.E., Leong, H., McEvoy, L., Whitfield, S., Du, R., Rush, G.: Monitoring working memory load during computer-based tasks with EEG pattern recognition methods. Hum. Factors J. Hum. Factors Ergon. Soc. **40**(1), 79–91 (1998)
3. Haapalainen, E., Kim, S., Forlizzi, J.F., Dey, A.K.: Psycho-physiological measures for assessing cognitive load. In: Proceedings of the 12th ACM international conference on Ubiquitous computing, pp. 301–310. ACM (2010)
4. Kim, J.: Bimodal emotion recognition using speech and physiological changes. Citeseer (2007)

5. Kim, J., André, E.: Emotion recognition based on physiological changes in music listening. IEEE Trans. Pattern Anal. Mach. Intell. **30**(12), 2067–2083 (2008)
6. Mayer, J.D., Geher, G.: Emotional intelligence and the identification of emotion. Intelligence **22**(2), 89–113 (1996)
7. Meudt, S., Zharkov, D., Kächele, M., Schwenker, F.: Multi classifier systems and forward backward feature selection algorithms to classify emotional coloured speech. In: Proceedings of the 15th ACM on International conference on multimodal interaction, pp. 551–556. ACM (2013)
8. Pantic, M., Rothkrantz, L.J.: Toward an affect-sensitive multimodal human-computer interaction. Proc. IEEE **91**(9), 1370–1390 (2003)
9. Phinyomark, A., Limsakul, C., Phukpattaranont, P.: A novel feature extraction for robust emg pattern recognition. arXiv preprint arXiv:0912.3973 (2009)
10. Picard, R.W.: Affective Computing, vol. 252. MIT Press, Cambridge (1997)
11. Scherer, K.R.: Speech and emotional states. In: Speech Evaluation in Psychiatry, pp. 189–220 (1981)
12. Schüssel, F., Honold, F., Bubalo, N., Huckauf, A., Traue, H., Hazer-Rau, D.: In-depth analysis of multimodal interaction: an explorative paradigm. In: Kurosu, M. (ed.) HCI 2016. LNCS, vol. 9732, pp. 233–240. Springer, Cham (2016). doi:10.1007/978-3-319-39516-6_22
13. Tang, X., Shu, L.: Classification of electrocardiogram signals with rs and quantum neural networks. Int. J. Multimedia Ubiquit. Eng. **9**(2), 363–372 (2014)
14. Tkach, D., Huang, H., Kuiken, T.A.: Study of stability of time-domain features for electromyographic pattern recognition. J. Neuroengineering Rehabil. **7**(1), 1 (2010)
15. Zhao, Q., Zhang, L.: ECG feature extraction and classification using wavelet transform and support vector machines. In: International Conference on Neural Networks and Brain, ICNN&B 2005, vol. 2, pp. 1089–1092. IEEE (2005)

Human Mobility-Pattern Discovery and Next-Place Prediction from GPS Data

Faina Khoroshevsky and Boaz Lerner$^{(\boxtimes)}$

Ben-Gurion University of the Negev, Beer-Sheva, Israel
{bordezki,boaz}@bgu.ac.il

Abstract. We provide a novel algorithm for the discovery of mobility patterns and prediction of users' destination locations, both in terms of geographic coordinates and semantic meaning. We did not use any semantic data voluntarily provided by a user, and there was no sharing of data among the users. An advantage of our algorithm is that it allows a trade-off between prediction accuracy and information. Experimental validation was conducted on a GPS dataset collected in the Microsoft Research Asia GeoLife project by 168 users in a period of over five years.

Keywords: GeoLife · Human behavior · Location extraction · Mobility pattern · Next place prediction · Positioning technology · Semantic information · Stay point · Trajectory data

1 Introduction

Positioning technologies, such as GPS, provide accurate and continuous geographical positions of mobile devices. Even though human movement and mobility patterns display a high degree of freedom and variation, they also exhibit structural patterns due to geographic and social constraints. Thus, the problem of accurately predicting a user destination location based on the user's mobility patterns becomes increasingly important in different areas, one of which is related to contextual applications.

One of two main ways to predict a next location is by predicting a semantic label [6,20], such as "Restaurant", without providing the coordinates of that label. This might be ambiguous since users visit multiple restaurants in different locations. In addition, explicit semantic information may not be available, or when it is, using it may compromise the user's privacy. The other way of prediction is of a specific geographical location ID with the corresponding coordinates, using only anonymous geographical data such as GPS or WiFi records [5,8,21]. The prediction of a location ID is a more difficult task because it requires a more precise output, which considers many more prediction options.

Existing studies on user location prediction can be classified into three categories: (1) those using only a user's own data, (2) those using the data generated by crowds, and (3) hybrid methods using both kinds of data [26]. For the task of next place prediction, one may benefit from using other users' data. Even though

© Springer International Publishing AG 2017
F. Schwenker and S. Scherer (Eds.): MPRSS 2016, LNAI 10183, pp. 24–35, 2017.
DOI: 10.1007/978-3-319-59259-6_3

the user has visited many locations, there must be some places the user has never been [26]. These places cannot be predicted, as opposed to a case in which the list of all possible locations to be visited is based on the crowd. A model that is based only on a user's own data may be considered as a personalized model, and a model that incorporates other users' data as a generalized model, and given observations from all users over the whole recording period, one could expect the general model to be robust prior to improving the performance of the personalized model [5]. Additionally, thanks to its user-independent nature, the general model can be used for new users without retraining [5].

On the other hand, using information regarding other users for the prediction task will force the use of a cloud service. In those cases, the calculations are usually performed on the server side where all data from all end users is aggregated and stored. While the on-line cloud service has become increasingly popular through the years, a user's privacy can be easily violated, as it is quite common for a cloud application provider to utilize user data for all sorts of claimed purposes, and it is nearly impossible for users to monitor the usage of their data [12], and they have to trust service providers with their personal data [12]. This is perhaps the biggest concern to the user community, and so far it has prevented many individual users and corporations from adopting cloud-based solutions. Therefore, storing the data and performing all calculations on the client side (user's device) will contribute to the user privacy, as it will prevent unnecessary personal identifiable information (PII) or sensitive location data transfer between the server and the client side. This was the main motivation in this study to develop an algorithm that avoids any data sharing among users.

Section 2 presents related work, whereas Sect. 3 introduces our proposed framework. Section 4 demonstrates our results before Sect. 5 concludes the work and also offers further research.

2 Related Work

Next places are commonly predicted using data-mining techniques based on collected mobility traces. In this work, we consider the prediction of one of the learned locations in terms of location ID. Others [10] presented a prediction model that represents the mobility behavior of an individual as a Markov model and predicts the next location based on the previously visited locations. They modified a similar approach [2] using a different clustering algorithm to find locations. A spatio-temporal approach based on nonlinear analysis of the time series of start times and duration times of visits was introduced in [24], and a trajectory model, which is represented as a probabilistic suffix tree with both spatial and temporal information of movements, was described in [15]. An algorithm that predicts a user's next place using a support vector machine classifier given only the current context of time stamp and location was presented in [21]. Several mobility predictors based on graphical models, neural networks, and decision trees were suggested [8], as well as their ensemble, similar in spirit to an ensemble method in which different mobility patterns were extracted with multiple models and combined under a probabilistic framework [5].

3   Framework

We present an algorithm that addresses the task of predicting users' destination locations without given semantic data by the user himself and without sharing data among users, and test it in predicting end locations of trajectories.

3.1   Database Description and Trajectory Collection

We used a GPS record dataset that was collected in the Microsoft Research Asia GeoLife project over a five-year period, from April 2007 to August 2012 [28,29,31]. We transformed the raw GeoLife trajectories to represent contiguous sequences of points having a "big enough" gap between them. The gaps between points are measured both in terms of time duration and distance with the thresholds set to be at least 20 min and 50 m, with the time constraint tested first. The GeoLife dataset originally contained 182 users, but we used only 168 users who had at least five trajectories; at least three trajectories for training and two for testing.

3.2   Stay-Point Detection

A stay point is a geographic region where the user stayed for a period of time, where time and distance can be considered as attributes to decide on a "stay" [17]. Once we have the sequence of start, stay, and end points, we obtain a meaningful trajectory, as opposed to the initial sequence of the raw GPS points.

3.3   Clustering Stay Points to Locations by Geographic Proximity

Finding users' locations using the raw GPS points has been previously performed by a variant of the k-means clustering algorithm [2], based on popular destinations, as those destinations where the users spend the most time [14], and using a density-based clustering algorithms, such as DBSCAN [7] and DJ-clustering [32], which use the density of local neighborhoods of points. Also others denoted locations as clusters of stay points [30] or proposed a grid-clustering algorithm that finds locations based on clustering stay points into locations of some fixed size [27]. The last method was further improved [5] by adding an additional step of maximizing the number of covered stay points with respect to the set of regions that cover the highest density unassigned cell.

   Our main purpose was to find these locations with as few a priori limitations as possible, e.g., without assuming a precise location radius [2] or area [5,27], or predefining the minimum number of points in a location [7]. Furthermore, we claim that different users should have the option to have different sizes of locations to best represent their daily life.

   Based on these considerations, we developed an algorithm that clusters stay points hierarchically [13] using the complete-linkage criterion and the Haversine distance [18]. We refer to the maximum distance between clusters at each clustering step as parameter H, and use the Davies-Bouldin (DB) index [4] and the Silhouette coefficient (SC) [23]—as well as two other metrics we developed (Sect. 3.5)—to evaluate the clustering and choose the optimal value of H.

3.4 Determining Semantic Meaning to Location

Knowing the semantic places around each point clustered in a location, we can give this location a semantic meaning. To do this, we used semantic data from the OSM project [1], using its application programming interface (API) implemented by the R software package "osmar" [9]. This approach is safer than using semantic information that is provided by the user himself (such as where his home or workplace is) in terms of data privacy. Also, it may be more accurate, as e.g., a shop, which is a possible semantic value to extract from a map, can be a place someone goes to in order to buy groceries, but it can also be a workplace. Thus, collecting a list of unique semantic values of all points in a location represents the semantics of that location.

3.5 Clustering Evaluation

To address our specific task of clustering geographical points to semantically meaningful locations, we developed two new clustering evaluation metrics that should be maximized. These are based on the DB and SC metrics in conjunction with a *semantic score* we have defined that is based on the (weighted by number of points) average semantic similarity between all pairs of points in a cluster (i.e., the fraction of common semantic values for these pairs):

1. 0.5[Normalized semantic score(H) − Normalized DB index(H)]
2. 0.5[Normalized semantic score(H) + Normalized SC(H)]

The four metrics evaluated the hierarchical clustering and recommended the optimal value of H. Each metric yielded its own best value of H, which results in a different number of clusters (locations) providing a different resolution.

3.6 Clustering Trajectories of Similar Patterns

To represent user's location mobility patterns, we want to detect groups of similar trajectories for the user. After clustering stay points to locations, we substitute [30] a start/stay/end point in a user's location history with the cluster ID the point pertains to, and thereby can represent this history as a sequence of visited locations [17]. Then, trajectories are represented as sequences of locations, i.e., string sequences. To cluster similar trajectories, we require a similarity/distance metric between string sequences [11, 16], and we applied a previously recommended one [22].

To compare sequences that may be of different lengths, we map a sequence x using an $|L|$-dimensional feature space of "language" sub sequences $w \in L$ by calculating an embedding function $\phi_w(x)$ for every w appearing in x, and then measure the similarity between equal-length vectors in the $|L|$-dimensional feature space. We chose the Blended k-grams method [25] for the formal language, where L is the set of all sub sequences of lengths 1 and 2, and the frequency of each $w \in L$ in each sequence is the embedding function. The normalized linear kernel (cosine function) was chosen as the similarity measure between pairs of trajectories, and thus the distance between trajectories $T1$ and $T2$ is defined as: $distance(T1, T2) = 1 - similarity(T1, T2)$.

To measure distances between trajectories that are represented as sequences of lists of semantic places, e.g.,

$T1 : [\text{``bank''}] \rightarrow [\text{``sport''}, \text{``shop''}] \rightarrow [\text{``transport''}, \text{``sport''}]$ and
$T2 : [\text{``transport''}] \rightarrow [\text{``shop''}, \text{``office''}]$,

we generalize the framework by establishing a language comprising sub sequences of both length 1, e.g., {"bank", "sport", "shop", "transport", "office"}, and length 2, e.g., {"bank_sport", "bank_shop", "sport_transport", "sport_sport", "shop_transport", "shop_sport", "transport_shop", "transport_office"}, to a unified list, L = { "bank", "sport", "shop", "transport", "office", "bank_sport", "bank_shop", "sport_transport", "sport_sport", "shop_transport", "shop_sport", "transport_shop", "transport_office"}.

Once the distance measurement between trajectories has been defined, it is possible to apply the hierarchical clustering algorithm also to trajectories (of both types), where the clustering results are evaluated using the DB index.

3.7   Preparing a New Trajectory for Prediction

To predict next location in a new (test) raw trajectory, we first assign the trajectory to the most representative trajectory cluster among those that were already found during training (Sect. 3.6). This is performed in the following way:

1. For each start/stay/end point in the new trajectory, check if allocating this point to an existing location will not violate the maximum distance restriction between pairs of points in that cluster (location) with a margin of 10%. Among all suitable clusters, choose the one with the minimal average distance between the test point and all training points of the cluster. If there is no suitable cluster, the location ID for the point will be "0", meaning "other location".
2. Represent the test trajectory as both a sequence of location IDs and semantic values (similar to training).
3. For each type of sequence representation (IDs and semantics), find for each test trajectory the closest group (cluster) of trajectories and assign this trajectory to that cluster.

3.8   Prediction of a Next Place

In this study, next place prediction is based on a history of visits for a user, a priori information that is independent of the user trajectories, or a combination of the two. We designed ten prediction functions, each of which utilizes another piece of information about location transitions the user makes or time and environment conditions for making their trajectory, and an eleventh function that combines the ten functions using a random forest (RF) [3]. Similar to [2,10], functions f1–f4 predict a next place for a new trajectory that was assigned a cluster using a location transition matrix for that cluster that measures the probabilities to transit from one location to another based on the seen transitions between the trajectories of that cluster. Functions f5–f10 assume that our visits at different locations are highly influenced by the day of the week, the time of day, or the proximity to other locations. In all functions, if there is more than one prediction option, for example, if there are equal transition probabilities to several locations, then we choose the closest one to the current location.

f1: "Predict by transition probabilities among locations". This function is based on the trajectory clusters derived during training when the trajectories are represented as sequences of location IDs. For each cluster of trajectories, a transition matrix was built according to the transitions among the locations of the trajectories of that specific cluster. Given a new trajectory, it is assigned to its most similar cluster of trajectories. The prediction of the end location of this test trajectory is based on the last known location of the user in that particular trajectory and on the highest transition probability from that location. If the location ID is found, but the probability of transition from it to every possible other location is 0, the prediction will be location number "0", meaning the prediction is "other" or an "unknown" location. If the location ID of the last known point is "0", we go back to one previous location and repeat the procedure. If eventually nothing is found, the prediction location number is "0", meaning "other".

f2: "Predict by transition probabilities among semantic locations". For each cluster of trajectories represented as semantic sequences, we build a transition matrix to describe the probabilities in moving from one semantic place to another. For each new trajectory, we predict the semantics of its end point according to the highest probability of transition from the previous point to the end point. If such prediction is possible (and does not result in "other"), then we choose the location ID that corresponds to this semantic location to represent this end of trajectory.

f3: "Choose f2 if f1 = 0". If the prediction output by f1 is "0", predict by f2. If the semantic prediction also results in "other", then this will be the final prediction. This heuristic combines the first two in a way that prioritizes f1 over f2. It does that because if f1 returns an answer which is not "other", then it results in a specific answer of a location ID in comparison to f2, which returns a more general result that may fit several locations.

f4: "Another semantic prediction". For each cluster of trajectories, which were clustered by semantic information, consider the transitions between all stay points in a location. For every point, choose the predicted place with the highest transition probability. The predicted semantic location will be the one that corresponds to the semantic place that was selected by most points.

f5: "Predict by day–1ˢᵗ option". This prediction function uses the information we can collect of the user's visits to each location by the day of the week. Based on the training trajectories, we calculate the conditional probability of a location given each day of the week. Given the user's current day of visit, we choose the prediction of the next location to be the location with the highest probability to visit on this day.

f6: "Predict by day–2ⁿᵈ option". Following the same steps as f5, instead of choosing the predicted location to be the one with the highest probability, consider all locations with a visiting probability >0 and choose the closest one to the current location.

f7: "Predict by the hour". This prediction function uses the information collected about the number of user's visits to each location grouped by hour of the day. In order to predict the next location, we make the following assumptions: (a) the time the user will arrive to his next location is unknown, (b) only the current time is known, and (c) the user will arrive to his next location at a close point in time to the current time. Given these assumptions, we look at the probabilities of being at each location, starting at the current hour, and search for the first non-zero probability. One would notice that these probabilities are incomparable due to different scales as a result of different probabilities to visit the location in the first place in any given 24-hour period. Therefore, the probabilities are weighted by the total probability of being at each location, regardless of the time, by counting the number of visits at each location and dividing by the total number of visits.

f8: "Predict the closest location". This is a very simple heuristic that is based on the assumption that sometimes people who need to be at several places (for some errands for example), and can choose in what order to visit those locations, will tend to move to the next closest location each time. So with this heuristic, the prediction of the next location is simply choosing the closest location to the current one.

f9: "Predict by day and hour–1ˢᵗ option". This prediction function uses the information we can collect on the user's visits to each location not only by day, and not only by hour, but by the combination of both. In order to be able to compare these probabilities, they are weighted by the total probability of being at each location regardless of the time or day. The result of the prediction is the location with the highest weighted probability.

f10: "Predict by day and hour–2ⁿᵈ option". This prediction function does the same as f9, but instead of choosing the prediction as the location with the highest weighted probability, it considers all options with a non-zero weighted probability, and among them chooses the closest location to the current one.

Since the ten functions aiming to predict the end location of a trajectory are heuristic, we designed a mechanism that combines their decisions by the RF classifier. The RF is trained to map each ten-dimensional vector of decisions (locations) to the location known in the training set. Then based on the ten-dimensional function representation of the end location in the test trajectory and the trained RF, the end location is predicted. We experimentally evaluated the RF with a range of 64 to 128 trees (similar to [19]), and found 100 trees provide good performance with only little sensitivity to this size.

4   Results and Discussion

To test our algorithm, we partitioned the set of trajectories of each user chronologically, such that the first 80% of the trajectories are used for training, and the remaining 20% for testing. We tested our algorithm by predicting the end location of each trajectory of the test set. We define the prediction accuracy of each user as the ratio between the number of correct predictions to the number of trajectories in his test set. Table 1 presents the results of the average (Avg) and standard deviation (Std) of the prediction accuracy over all users, as a function of the clustering evaluation metric.

Table 1. Prediction accuracy and Std by the clustering metric

Clustering by	Avg (Std) of prediction functions and RF										
	f1	f2	f3	f4	f5	f6	f7	f8	f9	f10	RF
DB	0.74	0.61	0.72	0.63	0.53	0.53	0.50	0.54	0.53	0.53	0.59
	(0.33)	(0.37)	(0.34)	(0.38)	(0.40)	(0.41)	(0.41)	(0.43)	(0.39)	(0.41)	(0.40)
DB+semantics	0.53	0.44	0.49	0.38	0.18	0.15	0.16	0.15	0.2	0.15	0.23
	(0.30)	(0.30)	(0.31)	(0.32)	(0.29)	(0.29)	(0.29)	(0.30)	(0.29)	(0.29)	(0.30)
SC+semantics	0.72	0.58	0.71	0.62	0.52	0.53	0.47	0.56	0.51	0.53	0.62
	(0.33)	(0.36)	(0.32)	(0.36)	(0.38)	(0.40)	(0.40)	(0.41)	(0.38)	(0.40)	(0.37)
SC	0.82	0.68	0.81	0.73	0.61	0.65	0.63	0.72	0.6	0.65	0.76
	(0.28)	(0.37)	(0.29)	(0.35)	(0.38)	(0.38)	(0.38)	(0.37)	(0.38)	(0.38)	(0.33)

We can see that for any clustering evaluation metric, predictor f1 results in the highest prediction accuracy, even above the RF predictions. Since the RF classifier cannot predict an "unknown" ("0") location (as these are not included in the training set), we define our final prediction as a combination of RF and the predictor f1 in the following way: if the prediction output of f1 for a given trajectory is "0", then our final prediction will be "0" as well, but in any other case, the final prediction will be the prediction of RF. Combining RF with f1 yields the best prediction accuracy, regardless of the clustering metric.

Table 2 compares our algorithm (first four rows) with algorithms of previous works, in terms of the average number of locations found per user, prediction accuracy, and main assumptions made by the algorithm. We compare ourselves only to works that predicted a location ID and that specify information regarding the number of locations found for a user. As can be seen, our results are on par with previously reported works. Not only that our algorithm provides

Table 2. Average prediction accuracy and number of locations needed

	Dataset	Number of locations	Accuracy	Assumptions	Personalized model (Yes/No)
SC	Geolife (168 users)	4	0.82	None	Yes
DB	Geolife (168 users)	19	0.75	None	Yes
SC+semantics	Geolife (168 users)	15	0.72	None	Yes
DB+semantics	Geolife (168 users)	93	0.54	None	Yes
Gambs et al. 2012	Phonetic dataset	5	0.7–0.95	DJ-Clustering algorithm assumptions	Yes
Gambs et al. 2012	Geolife (175 users)	8	0.7–0.95	DJ-Clustering algorithm assumptions	Yes
Do and Gatica-Perez, 2012	LDCC dataset	37	0.6	Location is 250 m × 250 m	Yes
Do and Gatica-Perez, 2012	LDCC dataset	37	0.64	Location is 250 m × 250 m	No
Prabhala et al. 2014	Nokia Mobility Data Challenge and WTD dataset	150–400	0.5	Locations are known	Yes
Etter et al. 2013	Nokia Mobility Data Challenge	150–400 [21]	0.56	Locations are known	Yes

a reasonable degree of accuracy, it also does it with a reasonable number of locations.

The different clustering evaluation metrics in this work result in different numbers of clusters with different sizes. Thus, we have defined a simple density measure of the number of points per unit of area. The measure is calculated assuming that each location (cluster) is a circle with a diameter equals to the distance between the two most distant points within the cluster. Hence, given $D_{max}{}^i$, the distance between the two most distant points in the ith cluster, and n^i, the number of points associated with the cluster, the cluster density, ρ_i, can be calculated as,

$$\rho_i = \frac{n^i}{\pi\left(\frac{D_{max}{}^i}{2}\right)^2} = \frac{4n^i}{\pi\left(D_{max}{}^i\right)^2}. \tag{1}$$

Since each clustering method produces multiple clusters for every user, we weight the cluster density by the number of points in it to create a single density measure per user.

Figure 1 presents the average over all users of the prediction accuracy vs. location (cluster) density for each clustering alternative. We can conclude that regardless of the clustering alternative (with the exception of SC), the accuracy results decrease for the users with the increased average density. That means that the evaluation metrics that generate more clusters, which are smaller and denser, provide us with more meaningful locations on the one hand, but on the other hand, introduce a higher variability in location history, thus making every

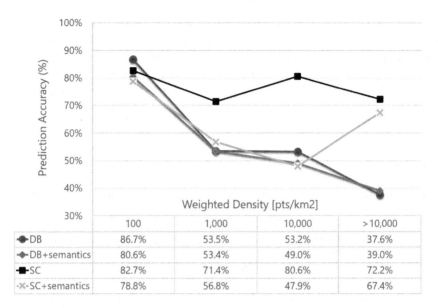

	100	1,000	10,000	>10,000
DB	86.7%	53.5%	53.2%	37.6%
DB+semantics	80.6%	53.4%	49.0%	39.0%
SC	82.7%	71.4%	80.6%	72.2%
SC+semantics	78.8%	56.8%	47.9%	67.4%

Fig. 1. Prediction accuracy for different clustering evaluation metrics and density ranges.

location more difficult to predict. A simple analogy to a real life scenario can be emphasized when one is trying to describe user location history, e.g., "visited retail store, then nearby restaurant, then post-office" rather than "visited shopping mall" where all these locations are. The implication of having a higher resolution, however, is that there are many more locations created by the clustering method, and this strongly affects the classification accuracy. Using the previous example, naturally, it is easier to predict that a user is going to visit the mall than the precise store or restaurant he is going to visit there.

5   Conclusion

The problem of accurately predicting a user's destination location is a difficult one to solve. Given that the user's trajectories are collected over a large period of time, they contain noise and tend to be incomplete. Predicting the user's next place accurately and consistently is a vexing problem [20].

This work is based on the assumption of no data sharing among users, and that a user's data is stored on his own device, in order to reduce the risks of invasion of privacy. Our work demonstrates that the predictability of user mobility is strongly related with the number and density of the users' locations, as learned from the data of each user. We did not decide a priori on a precise number of locations or on a fixed size of locations. We rather present the number of locations and their sizes as an output of our procedure of location extraction. The new proposed algorithm relaxes many assumptions used in other works in the field and demonstrates prediction accuracy in the range of 54%–82% for different location extraction approaches, which is on par with the reported values in similar studies. Future research may consider other clustering evaluation metrics. One may also consider using different similarity measures for the procedure of finding clusters of similar trajectories as presented in [22]. For the prediction algorithm, a different partitioning of the user's data into training and test sets can be evaluated.

References

1. Openstreetmap. http://wiki.openstreetmap.org/wiki/About. Accessed 01 Jan 2016
2. Ashbrook, D., Starner, T.: Using GPS to learn significant locations and predict movement across multiple users. Pers. Ubiquit. Comput. **7**(5), 275–286 (2003)
3. Breiman, L.: Random forests. Mach. Learn. **45**(1), 5–32 (2001)
4. Davies, D.L., Bouldin, D.W.: A cluster separation measure. IEEE Trans. Pattern Anal. Mach. Intell. **1**(2), 224–227 (1979)
5. Do, T.M.T., Gatica-Perez, D.: Contextual conditional models for smartphone-based human mobility prediction. In: Proceeding of the 2012 ACM Conference on Ubiquitous Computing, pp. 163–172 (2012)
6. Do, T.M.T., Gatica-Perez, D.: Where and what: using smartphones to predict next locations and applications in daily life. Pervasive Mob. Comput. **12**, 79–91 (2014)
7. Ester, M., Kriegel, H.P., Sander, J., Xu, X.: A density-based algorithm for discovering clusters in large spatial databases with noise. In: KDD, vol. 96, pp. 226–231 (1996)
8. Etter, V., Kafsi, M., Kazemi, E., Grossglauser, M., Thiran, P.: Where to go from here? Mobility prediction from instantaneous information. Pervasive Mob. Comput. **9**(6), 784–797 (2013)
9. Eugster, M.J.A., Schlesinger, T.: osmar: OpenStreetMap and R. The R Journal **5**(1), 53–63 (2013)
10. Gambs, S., Killijian, M.O., del Prado Cortez, M.N.: Next place prediction using mobility Markov chains. In: Proceeding of the 1st Workshop on Measurement, Privacy, and Mobility, p. 3. ACM (2012)
11. Hamming, R.W.: Error detecting and error correcting codes. Bell Syst. Tech. J. **29**(2), 147–160 (1950)
12. Hung, S.H., Shih, C.S., Shieh, J.P., Lee, C.P., Huang, Y.H.: Executing mobile applications on the cloud: framework and issues. Comput. Math. Appl. **63**(2), 573–587 (2012)
13. Johnson, S.C.: Hierarchical clustering schemes. Psychometrika **32**(3), 241–254 (1967)
14. Kim, M., Kotz, D., Kim, S.: Extracting a mobility model from real user traces. In: INFOCOM, vol. 6, pp. 1–13 (2006)
15. Lei, P.R., Shen, T.J., Peng, W.C., Su, J.: Exploring spatial-temporal trajectory model for location prediction. In: IEEE 12th International Conference on Mobile Data Management, vol. 1, pp. 58–67 (2011)
16. Levenshtein, V.I.: Binary codes capable of correcting deletions, insertions, and reversals. Sov. Phys. Dokl. **10**(8), 707–710 (1966)

17. Li, Q., Zheng, Y., Xie, X., Chen, Y., Liu, W., Ma, W.Y.: Mining user similarity based on location history. In: Proceeding of the 16th ACM SIGSPATIAL International Conference on Advances in Geographic Information Systems, no. 34. ACM (2008)

18. Mwemezi, J.J., Huang, Y.: Optimal facility location on spherical surfaces: algorithm and application. NY Sci. J. **4**(7), 21–28 (2011)

19. Oshiro, T.M., Perez, P.S., Baranauskas, J.A.: How many trees in a random forest? In: Perner, P. (ed.) MLDM 2012. LNCS, vol. 7376, pp. 154–168. Springer, Heidelberg (2012). doi:10.1007/978-3-642-31537-4_13

20. Prabhala, B., Porta, T.L.: Spatial and temporal considerations in next place predictions. In: IEEE Conference on Computer Communications Workshops (INFOCOM WKSHPS), pp. 390–395. IEEE (2015)

21. Prabhala, B., Wang, J., Deb, B., Porta, T.L., Han, J.: Leveraging periodicity in human mobility for next place prediction. In: IEEE Wireless Communications and Networking Conference (WCNC), pp. 2665–2670 (2014)

22. Rieck, K., Laskov, P.: Linear-time computation of similarity measures for sequential data. J. Mach. Learn. Res. **9**, 23–48 (2008)

23. Rousseeuw, P.J.: Silhouettes: a graphical aid to the interpretation and validation of cluster analysis. J. Comput. Appl. Math. **20**, 53–65 (1987)

24. Scellato, S., Musolesi, M., Mascolo, C., Latora, V., Campbell, A.T.: NextPlace: a spatio-temporal prediction framework for pervasive systems. In: Lyons, K., Hightower, J., Huang, E.M. (eds.) Pervasive 2011. LNCS, vol. 6696, pp. 152–169. Springer, Heidelberg (2011). doi:10.1007/978-3-642-21726-5_10

25. Shawe-Taylor, J., Cristianini, N.: Kernel Methods for Pattern Analysis. Cambridge University Press, Cambridge (2004)

26. Ying, J.J.C., Lee, W.C., Tseng, V.S.: Mining geographic-temporal-semantic patterns in trajectories for location prediction. ACM Trans. Intell. Syst. Technol. (TIST) **5**(1), 2:1–2:33 (2013)

27. Zheng, V.W., Zheng, Y., Xie, X., Yang, Q.: Collaborative location and activity recommendations with GPS history data. In: Proceeding of the 19th International Conference on the WWW, pp. 1029–1038. ACM (2010)

28. Zheng, Y., Li, Q., Chen, Y., Xie, X., Ma, W.Y.: Understanding mobility based on GPS data. In: Proceeding of the 10th International Conference on Ubiquitous Computing, pp. 312–321. ACM (2008)

29. Zheng, Y., Xie, X., Ma, W.Y.: GeoLife: a collaborative social networking service among user, location and trajectory. IEEE Data Eng. Bull. **33**(2), 312–321 (2010)

30. Zheng, Y., Zhang, L., Xie, X., Ma, W.Y.: Mining correlation between locations using human location history. In: Proceeding of the 17th ACM SIGSPATIAL International Conference on Advances in Geographic Information Systems, pp. 472–475. ACM (2009)

31. Zheng, Y., Zhang, L., Xie, X., Ma, W.Y.: Mining interesting locations and travel sequences from GPS trajectories. In: Proceeding of the 18th International Conference on WWW, pp. 791–800. ACM (2009)

32. Zhou, C., Frankowski, D., Ludford, P., Shekhar, S., Terveen, L.: Discovering personal gazetteers: an interactive clustering approach. In: Proceeding of the 12th Annual ACM International Workshop on Geographic Information Systems, pp. 266–273. ACM (2004)

Fusion Architectures for Multimodal Cognitive Load Recognition

Daniel Kindsvater, Sascha Meudt$^{(\boxtimes)}$, and Friedhelm Schwenker

Institute for Neural Information Processing, Ulm University, 89069 Ulm, Germany
{daniel.kindsvater,sascha.meudt,friedhelm.schwenker}@uni-ulm.de

Abstract. Knowledge about the users emotional state is important to achieve human like, natural Human Computer Interaction (HCI) in modern technical systems. Humans rely on implicit signals like body gestures and posture, vocal changes (e.g. pitch) and mimic expressions when communicating. We investigate the relation between them and human emotion, specifically when completing easy or difficult tasks. Additionally we include physiological data which also differ in changes of cognitive load. We focus on discriminating between mental overload and mental underload, which can e.g. be useful in an e-tutorial system. Mental underload is a new term used to describe the state a person is in when completing a dull or boring task. It will be shown how to select suited features, build uni modal classifiers which then are combined to a multimodal mental load estimation by the use of Markov Fusion Networks (MFN) and Kalman Filter Fusion (KFF).

1 Introduction

A fundamental part of human communication is noticing a change in the affective state of the conversational partner. Affective state refers to the experience of feelings or emotions. To elaborate on this more, consider the following scenario: A person is telling another about a rather complex topic, e.g. in an teacher-student setting. During this conversation the student starts to look a bit overwhelmed by all the new information. In this case one would expect the teacher to change his pace as the student obviously can't follow up. Let that state the student is experiencing henceforth be referred to as *mental overload*. This term is meant to describe the state one is in when being confronted with a very complex task, e.g. understanding something completely new. The opposite, i.e. completing an easy task or listening to a teacher talking about a already well known topic, shall be called *mental underload*. In terms of the student-teacher example one can consider a electronic tutorial platform which controls its pace depending on the student's behavior. A user centered system should offer possibilities for the user to express their emotions [9].

In the past there had been a lot of research on detecting mental states utilizing single modalities, like detecting affect from loudness and pitch of words being sad [8,10]. Non-verbal ways of expressing and detecting feelings can be facial

F. Schwenker and S. Scherer (Eds.): MPRSS 2016, LNAI 10183, pp. 36–47, 2017.
DOI: 10.1007/978-3-319-59259-6_4

expressions [11,14] and gestures [7]. Furthermore a lot of studies conducted multimodal approaches on affective state estimation [12]. Most of them are based on acted or even strong expressive emotional datasets on basic emotions for example defined by Eckman [1]. This methods can not be easily transferred to typical weak expressive behaviors in HCI which in addition typically not refer to basic emotions. In this work we focus on the estimation of cognitive load called mental over and underload estimated from several modalities combined by using either a Markov Fusion Network or a Kalman Filter Fusion scheme. The utilized dataset contains data from natural behaving, none acting, users interacting with a HCI system which is able to induce different mental load levels.

2   Experimental Setting

The dataset is based on an experiment done within the Transregional Collaborative Research Centre SFB/TRR 62 "Companion-Technology for Cognitive Technical Systems".

Participants were asked to play a series of games based on the Interaction paradigm of Schüssel et al. [13]. The task of each game sequence was to identify the singleton element, i.e. the one item that is unique in shape and color (Number 36 and 2 in Fig. 1). The difficulty was set by adjusting the number of shapes shown and the time to answer. If the given answer was incorrect, they received no reward for that particular round. After a introduction each participant completed four game sequences of decreasing difficulty. The first sequence was designed to induce mental overload by a 6 × 6 board with 6 s to answer (see Fig. 1 left), the second was a 5 × 5 matrix with 10 s count down, the third was set to 3 × 3 size with 100 s, sequence four again was 3 × 3 mode with 100 s time to induce mental underload. A fifth sequence induces frustration, e.g. by purposely logging in a wrong answer. As the sequences one and four are explicitly designed to cause mental over- and underload, we focused only on those two.

After each sequence played the participants answered a Self Assessment Scale questionnaire (SAM). The aim of those questions was to determine valence, arousal and dominance experienced in the particular sequence. A total of 38

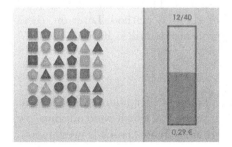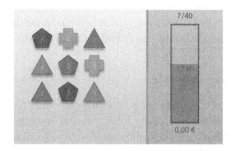

Fig. 1. Screen shot of the difficult level (left) with target element 36 and the easyest one (right) with target element 2.

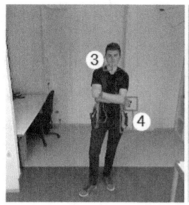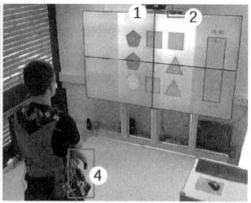

Fig. 2. Overview of the setting with sensors: (1) MS Kinect v2, (2) frontal webcam, (3) wireless headset, (4) GTec g.MOBIlab+ physiologic sensor with sensors attached to the users body.

participants were recorded. Of those were 18 male and 20 female. Their age spanned from 17 to 27 (mean 21.66, $\sigma^2 \approx 2.7$). For details on the Experimental setup and the usage of SAM to proof induction quality see [5].

During the experiment, participants were monitored by several sensors providing multimodal synchronous data. See Fig. 2. The sensory system contains two webcams (Logitech C9100), one in front locking towards the users face and one from the rear providing an overview of the scene, a wireless headset, a Microsoft Kinect v2 camera in front recording RGB, infrared, depth, skeleton/postural and acoustic information and finally a GTec g.MOBIlab+ biophysical sensor recorded Electrocardiography (ECG), Electromyography (EMG) of trapezius muscle, Electrodermal activity (EDA), respiration and body temperature.

3   Unimodal Recognition of Mental Overload

At first, six unimodal classifiers for three modalities were trained. One for speech, one for gestures and four for physiological signals. These unimodal classifiers are evaluated using the leave one subject out (LOSO) method. Later on, these classifiers are combined using the fusion approaches described in Sect. 4.

3.1   Speech

For cognitive load recognition from speech, a k-nearest neighbours classifier was trained. In total, 2067 training samples were presented which were obtained by extracting the following features from audio chunks where speech is present, i.e., a spoken word or sentence:

– Vocal pitch features: Local minima and maxima, first and second derivations

– Vocal energy features: Local minima and maxima, first and second derivations
– Mel-frequency cepstral coefficients (MFCC) features
– Duration related features: Length of the whole speech segment
– Frequency spectrum features: The distance between the 10% and 90% frequency quantile
– Harmonics-to-noise (HNR) features
– Voice quality features: Number of glottal pulses and jitter

These features are extracted by using the EmoVoice component of Social Signal Interpretation framework (SSI), which was proposed by Vogt et al. [15]. The work of Vogt et al. also provides a detailed description of these features.

A LOSO evaluation of this classifier yielded an accuracy of 62.26%.

3.2   Gesture

3247 samples (feature vectors) were used to train a k-nearest neighbours classifier for recognizing cognitive load from gestures. The data for feature extraction is provided by a Microsoft Kinect v2 sensor and contains the position of several body joints. The following body joints are used for feature extraction: Neck, left elbow, right elbow, left hand, right hand, right foot and left foot. For each of these joints the following features are computed: Mean velocity, variance of the velocity, mean acceleration and variance of the acceleration. Apart from that, the distances between the following joints are computed:

– Distance between hands
– Distance between right hand and head or left hand and head (the shorter one is chosen)
– Distance between right hand and left elbow
– Distance between left hand and right elbow
– Distance between feets
– Distance between right hand and right hip
– Distance between left hand and left hip

These features are computed by the SSI component EmoGesture, which was proposed by Hihn [5]. The work of Hihn also provides a detailed analysis of cognitive load estimation on the basis of gestures and postural behaviour.

Each feature vector (training sample) is computed on the basis of a four seconds data chunk. Only samples which hold enough gestural activity were used to train the classifier. Enough gestural activity is given if the mean velocity of at least one body joint overcomes a certain threshold. The classifier was evaluated using the LOSO method. An accuracy of 58.39% was obtained.

3.3   Physiology

In order to recognize mental overload on the basis of physiology, a k-nearest neighbours classifier for each of the following physiological signals was trained: ECG, EMG, EDA and body temperature. For EDA, EMG and temperature fifty statistical features are extracted, such as mean value, first and second derivative, standard deviation, local maxima and mean value of all local maxima, just to name some of them. These features are extracted on five seconds data chunks with an overlap of 4.5 s. The extraction is accomplished by an SSI component which is developed and implemented by Held [4]. The work of Held provides a detailed description of these features and a detailed analysis of cognitive load estimation on the basis of physiological signals and speech. For the ECG signals the following features are extracted:

- Mean interval between two R deflections in milliseconds (also called RR interval) (R deflection denotes the major deflection in an ECG signal)
- Mean heart rate in beats per minute
- Standard deviation of the first feature
- Standard deviation of the second feature
- Coefficient of variation of the first feature
- Root mean square of the difference between all successive RR intervals
- Number of pairs of RR intervals differing by 20 ms and more in %
- Number of pairs of RR intervals differing by 50 ms and more in %

In contrast to the other physiological signals, the features for the ECG signals are extracted on a 7.5 s time frame with an overlap of 6.5 s.

All four classifiers were evaluated using the LOSO method. Table 1 shows the number of presented samples and the obtained accuracies. The number of presented samples is much higher for physiological signals than for speech and gestures because they are always present and no activity detection is needed for this modality.

Table 1. Evaluation results of all physiological classifiers using the LOSO method.

Physiological signal	Number of samples	Accuracy
ECG	19928	50.04%
EMG	44601	56.27%
EDA	44601	52.55%
Temperature	44601	54.14%

4   Multimodal Recognition of Mental Overload

In order to obtain a higher overall accuracy, three fusion approaches were used to aggregate the above described classifiers. Since these fusion approaches have been implemented to work under real time conditions with real sensors, a suitable procedure for evaluating these fusion architectures must be applied. At first, a live performance of the whole system is simulated by using the recorded data of a subject, which wasn't used to train the classifiers, as input for the live system.

Only the data of the overload and underload sequence is used as input. The system works under real time conditions as if it gets data from actual sensors. The whole process from feature extraction to classifier fusion is applied just on time and the classifier decisions as well as the overall estimate of the MFN are written into a file. This process was repeated for twelve subjects. In the following the obtained data is referenced as *live performance data*.

4.1   Markov Fusion Network

The MFN proposed by Glodek et al. fuses multiple continuous classifier decisions in a certain time frame [3]. The MFN is organised in discrete time steps. Occurring unimodal classifier decisions are synchronized by assigning them to a discrete time step of the MFN. Finally, the MFN provides an overall estimate for each discrete time step. Furthermore, the MFN is capable to handle missing classifier decisions, i.e., the overall estimate y_t is always available even if there is no classifier decision for time step t. This is ensured by a Markov chain which enforces the estimates y_t to be similar to the estimates y_{t-1} and y_{t+1} which are close in time. Generally, y_t is influenced by the classifier decisions which are assigned to time step t and its temporal neighbours y_{t-1} and y_{t+1}.

The MFN is defined through three potentials, namely, *data potential, smoothness potential* and *distribution potential*. The *data potential* ensures that the overall estimates of the MFN are close to the unimodal classifier decisions. The *smoothness potential* ensures that the MFN estimates y_t are similar to their temporal neighbours y_{t-1} and y_{t+1}. The *distribution potential* ensures that the overall estimates of all classes sum up to 1 for each time step.

All three potentials are combined into one probability density function $p(\mathbf{Y}, \mathbf{X}_1, \ldots, \mathbf{X}_M)$, where $\mathbf{Y}$ are all estimates of the MFN, $\mathbf{X}_1, \ldots, \mathbf{X}_M$ are the unimodal classifier decisions and M the number of modalities. In order to obtain the most likely estimate $\mathbf{Y}$, only the mode of the probability density function has to be determined. This can be accomplished by gradient ascent. All gradients $\frac{\partial \mathbf{Y}}{\partial y_{i,t}}$ are computed, where i denotes the class and t the time step and then the estimates $y_{i,t}$ are modified towards the mode of the probability density function. This process is repeated until a certain number of iterations is reached. For the initialization of $\mathbf{Y}$ the mean of all classifier decisions or the least informative outcome (0.5 in case of a two-class problem) can be used. For *offline* fusion all time steps are taken into account at once, whereas for *online* fusion a sliding time window is implemented. A more detailed description of the MFN can be found in [3].

Numerical Evaluation. Figure 3 illustrates the *live performance data* of the MFN for subject 3 and 17. The x-axis denotes the discrete time step of the MFN. One time step represents 500 ms during live performance. The y-axis denotes the degree of mental overload. The blue dashed line represents the desired outcome, i.e., it indicates the overload and underload sequence. The green, blue, yellow, cyan, magenta and red curves represent (in the same order) the Kinect, EMG,

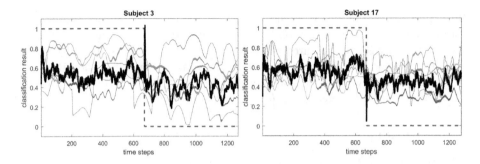

Fig. 3. Live performance of the MFN for subject 3 and 17. The green, blue, yellow, cyan, magenta and red curves represent (in the same order) the Kinect, EMG, EDA, temperature, ECG and audio classification results. The black curve shows the MFN outcome and the blue dashed line represents the desired outcome. (Color figure online)

EDA, temperature, ECG and audio classification results. For subject 17, the EDA (yellow curve), ECG (magenta curve) and temperature (cyan curve) classifiers don't perform very well. These curves seem to have no tendency. On the other hand, the Kinect (green curve), audio (red curve) and EMG (blue curve) classifiers tend to have higher outcomes in the overload sequence than in the underload sequence. The overall outcome of the live system (black curve) inherits this tendency and therefore the overall performance for subject 17 is very well.

Comparisons of the live performance for subject 17 with live performances of other subjects showed that every unimodal classifier performs different on different subjects. For example, the EDA classifier performs very well for subject 3 but poorly for subject 17, while the EMG classifier performs better for subject 17 than for subject 3. Furthermore, the unimodal outcomes are unstable and show strong oscillations. The fusion seems to provide a more stable and reliable overall outcome.

In order to compare the MFN with the unimodal classifiers and other fusion approaches, a measure for accuracy must be defined. One method to obtain an accuracy for the MFN is to set all outcomes above a certain threshold to 1 and all outcomes below this threshold to 0. Then, for every time step of a subjects live performance the MFN outcome is checked whether it is the desired value or not and an accuracy for a certain subject is obtained. Since the live performance was simulated for twelve subjects, an overall accuracy can be obtained by averaging all twelve subject specific accuracies. The black curve in Fig. 4 shows the accuracy (averaged over twelve subjects) of the MFN dependent on the threshold. The orange circles represent the optimal threshold and the corresponding achieved accuracy for each of the twelve individual subjects. The green and red curves show the accuracy for subject 19 and 29 dependent on their specific threshold.

The black curve shows that the default threshold of 0.5 is a suitable threshold in the general case. An average accuracy of 64.42% is achieved using this threshold. However, the orange circles show that the accuracy for many subjects

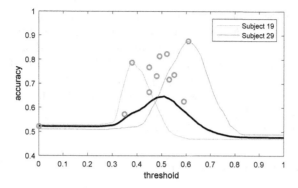

Fig. 4. MFN accuracy dependent on the threshold. The black curve shows the accuracy (averaged over twelve subjects) of the MFN dependent on the threshold. The orange circles represent the optimal threshold and the corresponding achieved accuracy for each of the twelve subjects. The green and red curves show the accuracy for subject 19 and 29 dependent on the threshold. (Color figure online)

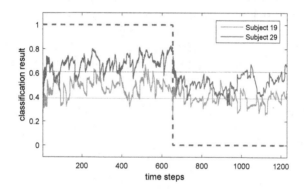

Fig. 5. Visualization of the MFN outcome for subject 19 and 29 by comparison and their individual optimal thresholds. The orange curve shows the overall outcome for subject 19 and the blue curve the overall outcome for subject 29. The blue dashed line represents the desired outcome. The green line shows the optimal threshold for subject 19 and the red one the optimal threshold for subject 29. (Color figure online)

can be improved if individualized thresholds are used. As illustrated by the green and red curves, for subject 19 and 29 the accuracy can be improved greatly if individualized thresholds are used instead of the default threshold of 0.5. This circumstance is also illustrated in Fig. 5. It depicts the live performance of the MFN for the subjects 19 and 29 by comparison. The orange curve shows the overall outcome for subject 19 and the blue curve the overall outcome for subject 29. The blue dashed line represents the desired outcome. The green line is the optimal threshold for subject 19 and the red one the optimal threshold for subject 29.

The outcome for subject 29 is higher than the outcome for subject 19 for almost every time step. These subjects seem to have different individual baselines on dealing with mental load, and therefore individual thresholds lead to much better accuracies. The accuracy averaged over all twelve subjects is 71.95% if individualized thresholds are used. Compared to the value of 64.42% it is a gain of 7.53%. This leads to the question how an individual threshold during live performance can be estimated. This could be accomplished by implementing something similar to a moving average filter.

4.2 Kalman Filter Based Fusion

Typically, Kalman filters [6] are used to fuse multiple measurements in order to obtain a more precise estimate than relying on just one measurement. In other words, noise and inaccuracies in measurements are reduced. Kalman filters are often used in object tracking, e.g., estimating the position of a plane. They can also be used to fuse multiple continuous classifier decisions as proposed by Glodek et al. [2]. In the case of Kalman filter based fusion (KFF), classifier outputs are interpreted as measurements. All measurements and the latest estimate of the Kalman filter are used to compute the next overall estimate. The estimation consists of two steps, namely prediction and update step. In the prediction step, the latest outcome is used to predict the outcome of the next time step. In the update step, the prediction is updated using the unimodal classifier decisions. Additionally, in both steps uncertainties are taken into account. Analogous to the traditional Kalman filter, these uncertainties are modelled as noise, i.e., additional parameters. Furthermore, the KFF is also capable of handling missing classifier decisions in a particular time step. A more detailed description of the KFF can be found in [2].

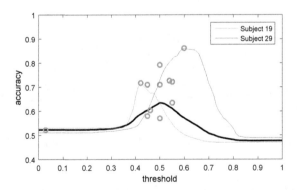

Fig. 6. KFF accuracy dependent on the threshold. The black curve shows the accuracy (averaged over twelve subjects) of the KFF dependent on the threshold. The orange circles represent the optimal threshold and the corresponding achieved accuracy for each of the twelve subjects. The green and red curves show the accuracy for subject 19 and 29 dependent on the threshold. (Color figure online)

Numerical Evaluation. In order to enable a comparison, the accuracy for the KFF approach was estimated the same way as the accuracy for the MFN as described above. Analogous to Figs. 4 and 6 shows the accuracy of the KFF dependent on the utilized threshold. The denotations are the same. The black curve shows that the default threshold of 0.5 is optimal in the general case, using this threshold an averaged accuracy of 63.50% is achieved. Similar to the MFN, the accuracy can be improved if individualized thresholds are used. However, the gain of accuracy is not as big as for the MFN. Firstly, the orange circles lie below the ones in Fig. 4. Secondly, the green and the red curve show that the KFF achieves a better individual accuracy for subject 19 and 29 than the MFN if the default threshold of 0.5 is used and therefore a smaller gain can be expected. An averaged accuracy of 67.87% and hence a gain of 4.37% is achieved if individualized thresholds are used.

4.3   Weighted Majority Voting

In order to enable a comparison of the MFN and the KFF to a classical decision aggregation method, the weighted majority voting was applied on the recorded classifier decisions of the *live performance data*. For each time step the available decisions were aggregated and afterwards an accuracy is obtained. The overall accuracy (averaged over twelve subjects) is 55.40% and lies significantly below the accuracy of the MFN and the KFF.

4.4   Discussion

The analysis of the *live performance data* and the LOSO evaluation results of the unimodal classifiers show that the unimodal classifiers are very unstable and unreliable. This instability is intensified by missing classifier decisions. Especially the audio classifier provides decisions very infrequently compared to the other classifiers. Furthermore, some classifiers perform well for one subject but bad for another subject, while other classifiers perform the other way around. The fusion via MFN and Kalman filter stabilises the overall outcome in such a way that strong oscillations are avoided and a more reliable outcome is provided even if many missing classifier decisions occur.

Furthermore, the accuracies of the MFN, the KFF and the weighted majority voting have been estimated. Tables 2 and 3 lists the accuracies of the unimodal classifiers and all tested fusion approaches. The MFN and the KFF perform slightly better than the best unimodal classifier. If individualized thresholds are used, the accuracies of these fusion approaches increase decently. The weighted majority voting is clearly outperformed by both fusion approaches.

These results show that the MFN and the KFF are both suitable fusion approaches when it comes to multimodal classifier fusion. The accuracy of the MFN is only 0.92% above the accuracy of the KFF, but if individual thresholds are used, the gap raises to 4.08%. One reason for the better performance of

Table 2. Achieved accuracies of the unimodal classifiers.

Modality	Accuracy
Kinect	58.39%
EMG	56.27%
EDA	52.55%
Temperature	54.14%
ECG	50.04%
Audio	62.26%

Table 3. Achieved accuracies of the fusion approaches.

Fusion approach	Accuracy
MFN	64.42%
KFF	63.50%
MFN (individual thresholds)	71.95%
KFF (individual thresholds)	67.87%
Weighted majority voting	55.40%

the MFN may be that the KFF takes only two time steps into account in its calculations, while the MFN takes time steps according to its window size into account.

Acknowledgments. The authors of this paper are partially funded by the Transregional Collaborative Research Centre SFB/TRR 62 "Companion-Technology for Cognitive Technical Systems" funded by the German Research Foundation (DFG).

References

1. Ekman, P.: An argument for basic emotions. Cogn. Emot. **6**(3–4), 169–200 (1992)
2. Glodek, M., Reuter, S., Schels, M., Dietmayer, K., Schwenker, F.: Kalman filter based classifier fusion for affective state recognition. In: Zhou, Z.H., Roli, F., Kittler, J. (eds.) MCS 2013. LNCS, vol. 7872, pp. 85–94. Springer, Heidelberg (2013). doi:10.1007/978-3-642-38067-9_8
3. Glodek, M., Schels, M., Schwenker, F., Palm, G.: Combination of sequential class distributions from multiple channels using Markov fusion networks. J. Multimodal User Interfaces **8**(3), 257–272 (2014)
4. Held, D.: Bimodale Erkennung affektiver Zustnde durch Ensemble Methoden anhand von Audio- und Biosignalen. Master's thesis, Ulm University (2016)
5. Hihn, H., Meudt, S., Schwenker, F.: Inferring mental overload based on postural behavior and gestures. In: Proceedings of the 2nd workshop on Emotion Representations and Modelling for Companion Systems. ACM (2016)
6. Kalman, R.E.: A new approach to linear filtering and prediction problems. J. Basic Eng. **82**(1), 35–45 (1960)
7. Kipp, M., Martin, J.-C.: Gesture and emotion: can basic gestural form features discriminate emotions? In: 3rd International Conference on Affective Computing and Intelligent Interaction Workshops, pp. 1–8. IEEE (2009)
8. Meudt, S., Zharkov, D., Kächele, M., Schwenker, F.: Multi classifier systems and forward backward feature selection algorithms to classify emotional coloured speech. In: Proceedings of the 15th ACM on International Conference on Multimodal Interaction, pp. 551–556. ACM (2013)
9. Picard, R.W.: Affective Computing, vol. 252. MIT Press, Cambridge (1997)

10. Plesa-Skwerer, D., Faja, S., Schofield, C., Verbalis, A., Tager-Flusberg, H., Dykens, E.M.: Perceiving facial and vocal expressions of emotion in individuals with Williams syndrome. Am. J. Ment. Retard. **111**(1), 15–26 (2006)
11. Russell, J.A., Bachorowski, J.-A., Fernández-Dols, J.-M.: Facial and vocal expressions of emotion. Annu. Rev. Psychol. **54**(1), 329–349 (2003)
12. Schels, M., Glodek, M., Meudt, S., Scherer, S., Schmidt, M., Layher, G., Tschechne, S., Brosch, T., Hrabal, D., Walter, S., et al. Multi-modal classifier-fusion for the recognition of emotions. In: Coverbal Synchrony in Human-Machine Interaction (2013)
13. Schüssel, F., Honold, F., Bubalo, N., Huckauf, A., Traue, H., Hazer-Rau, D.: In-depth analysis of multimodal interaction: an explorative paradigm. In: Kurosu, M. (ed.) HCI 2016. LNCS, vol. 9732, pp. 233–240. Springer, Cham (2016). doi:10.1007/978-3-319-39516-6_22
14. Shan, C., Gong, S., McOwan, P.W.: Robust facial expression recognition using local binary patterns. In: IEEE International Conference on Image Processing, 2005. ICIP 2005, vol. 2, p. II-370. IEEE (2005)
15. Vogt, T., André, E., Bee, N.: EmoVoice — a framework for online recognition of emotions from voice. In: André, E., Dybkjær, L., Minker, W., Neumann, H., Pieraccini, R., Weber, M. (eds.) PIT 2008. LNCS, vol. 5078, pp. 188–199. Springer, Heidelberg (2008). doi:10.1007/978-3-540-69369-7_21

Face Recognition in Home Security System Using Tensor Decomposition Based on Radix-(2 × 2) Hierarchical SVD

Roumen Kountchev[1], Suzan Anwar[2], Roumiana Kountcheva[3],
and Mariofanna Milanova[2(✉)]

[1] Technical University of Sofia, Sofia, Bulgaria
rkountch@tu-sofia.bg
[2] University of Arkansas at Little Rock, Little Rock, USA
{sxanwar,mgmilanova}@ualr.edu
[3] T&K Engineering, Sofia, Bulgaria
kountcheva_r@yahoo.com

Abstract. This paper explains research based on improving real time face recognition system using new Radix-(2 × 2) Hierarchical Singular Value Decomposition (HSVD) for 3rd order tensor. The scientific interest, aimed at the processing of image sequences represented as tensors, was significantly increased in the last years. Current home security solutions can be cost-prohibitive, prone to false alarms, and fail to alert the user of a break-in while they are away from the home. Because of advancements in facial detection and recognition techniques made in the past decade, we propose a home security system that takes advantage of this technology. To create such a system at a low cost requires algorithms that are powerful enough to detect users in various environmental conditions and fast enough to process real time video on weaker hardware. Experiments comparing the efficiency of two different decomposition techniques applied for face recognition in real time.

1 Introduction

Machine facial recognition is one of the staple problems in the field of computer vision and has been studied in great detail since the first experiments by Woodrow Bledsoe in the 1960's. Human beings are able to easily recognize faces at a young age thanks to specialized sections of our brains, but what seems like an easy task for humans is a very difficult task for computers.

When people talk about facial recognition, they often refer to a process that is actually performed in two separate steps, facial detection and facial recognition. First, regions that contain faces must be extracted from an image by recognizing the features that typically make up a face. After the face regions have been identified, the features of the unknown faces are compared to the features of known faces to determine their identity. Decomposing an image into its features is a task that is performed by both computers and our brains to make sense of the images that we are given. Many different algorithms exist for detecting a face in an image, and each algorithm has its own strengths and weaknesses depending on the context in which it is used. Likewise, there are several algorithms for recognizing a face and each has its uses.

© Springer International Publishing AG 2017
F. Schwenker and S. Scherer (Eds.): MPRSS 2016, LNAI 10183, pp. 48–59, 2017.
DOI: 10.1007/978-3-319-59259-6_5

In order to make a truly useful facial recognition system, faces must be detected in different lighting conditions and at various angles; simply making a system that has good results for recognizing frontal faces in controlled lighting conditions is not good enough for use in home surveillance. A home surveillance system must be able to work in low lighting conditions as many break-ins happen at night, and it must be able to detect faces from different angles because there is no guarantee that an intruder will be facing directly towards the camera at any point.

2 Related Work

Currently, the performance of face recognition algorithms increased implementing Convolution Neural Network (CNN). The power of CNN to extract knowledge from data has been verified in many fields [1]. Deep CNN is becoming the mainstream in face recognition [2]. There are many configurations of CNN but in all of them convolution nets use and process images as tensors.

Many of the most powerful face detection and recognition algorithms that achieve a high success rate in unconstrained conditions have already been commercialized, such as Picasa. Simplicam [3] has developed and commercialized a product that is very similar to the one described in this paper and is marketed towards people who want better security in their homes.

The basic methods for tensor decomposition [4] are higher-order extensions of the matrix SVD: the CANDECOMP/PARAFAC (CP) which decomposes the tensor as a sum of rank-one tensors, and the Tucker decomposition, which is a higher-order form of the Principal Component Analysis (PCA). To enhance the decomposition of image sequences (for example, human faces), represented as tensors, in this paper is proposed to use new approach, called Radix-(2×2) Hierarchical SVD (HSVD) [5, 6], which to replace famous Multilinear Singular Value Decomposition (MSVD) [7]. Proposed Radix-(2×2) HSVD is different from HSVD presented in [11]. In Radix-(2×2) HSVD tensor size $N \times N \times N$ for $N = 2^n$ is build from elementary tensor (ET) of size $2 \times 2 \times 2$.

The contributions of this paper are summarized as follows:

- We propose new tensor decomposition called 3D Hierarchical Singular Value Decomposition (3D HSVD)
- We implement 3D HSVD and developed Home Security System
- We compare two different algorithm for face recognition: face recognition using SVD and face recognition using 3D HSVD
- The main advantages of the new decomposition for video are the low computational complexity and the tree-like structure.

3 Proposed System

The developed home surveillance system is able to perform the following tasks:

- Generate TensorFaces from a database of facial images
- Monitor an area for motion

- Detect faces in frames where there is motion
- Recognize faces using the TensorFaces data compiled in step 1
- Send an SMS alert to the homeowner with a screenshot of the face and the name of the subject if the subject is recognized

A visual representation of the system's flow can be found below in Fig. 1.

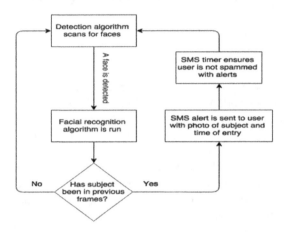

Fig. 1. Overall design

4   Face Detection

The field of face detection has made significant progress in the past decade, however there are still difficulties that come from lighting conditions, variations in scale, facial expressions, occlusions, etc. Generally speaking, face detection techniques can be divided into four main categories: *knowledge-based methods, feature-based approach, template matching methods and appearance-based methods* [12]. Appearance based methods have been shown to have better performance than the other methods. An appearance-based method learns face models from a set of representative training face images; the two important considerations in this process are which features to extract and which learning algorithm to use.

The Viola-Jones (VJ) algorithm is robust, works in real time video, and can detect faces and eyes without recognizing them. The first step in detecting a face using the VJ algorithm is to convert an image to grayscale. All N × N subsets of the grayscale image (where N is smaller than the size of will be checked to see if they match a set of Haar-like features (referred to as a Haar cascade). If several neighboring subsets match the Haar cascade, the region will be identified as a face. Faces of different sizes can be detected by scaling N up by an arbitrary multiplier and repeating the process. The VJ method of face detection proposes three main ideas that reduce computation time and make it possible for real time object detection: the integral image, the boosting algorithm, and the attention cascade structure [8].

The use of Haar-like features and their usage in face detection is well documented; an implementation of object detection using Haar-cascades can be found in the popular Open Source Computer Vision library (OpenCV) [13].

5 Tensor Decomposition Based on the Radix (2 × 2) Hierarchical Singular Value Decomposition

The basic idea is to represent the tensor decomposition of size $N \times N \times N$ (for $N = 2^n$), by using a hierarchical tree-like structure of N levels, consisting of elementary tensors of size $2 \times 2 \times 2$. In the first level each elementary tensor, which builds the tensor of size $N \times N \times N$, is decomposed by using SVD, and the so obtained components are rearranged in accordance with their energy. In the next levels, the edges of the small cube, which describes the elementary tensor, are increased twice. Each enlarged elementary tensor is decomposed once more through SVD; the new components are rearranged, etc. In brief, the heart of the method is given below.

1. Calculation of the SVD for elementary tensor of size 2 × 2 × 2

The tensor of size $2 \times 2 \times 2$, noted as $\mathbf{T}_{2\times2\times2}$, is the kernel of the decomposition for the tensor of size $N \times N \times N$ for $N = 2^n$. The HSVD algorithm for the tensor [T] of size $2 \times 2 \times 2$ (HSVD$_{2\times2\times2}$), based on the SVD$_{2\times2}$, is shown on Fig. 2 [5].

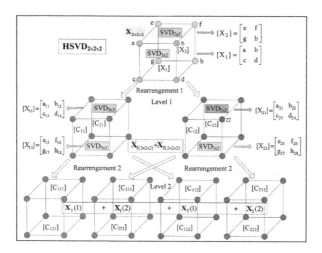

Fig. 2. Two-level HSVD$_{2\times2\times2}$ for elementary tensor of size $2 \times 2 \times 2$, based on the SVD$_{2\times2}$

After mode-1 unfolding the tensor $\mathbf{T}_{2\times2\times2}$, is obtained:

$$\text{unfold } \mathbf{T}_{2\times2\times2} = \begin{bmatrix} a & b & e & f \\ c & d & g & h \end{bmatrix} = \begin{bmatrix} [X_1] & [X_2] \end{bmatrix} \tag{1}$$

In the first level of $HSVD_{2\times2\times2}$, over each of the matrices $[X_1]$ and $[X_2]$ of the tensor $T_{2\times2\times2}$ is applied SVD of size 2×2 ($SVD_{2\times2}$) and in result is got:

$$[X_1] = \begin{bmatrix} a & b \\ c & d \end{bmatrix} = [C_{11}] + [C_{12}], \ [X_2] = \begin{bmatrix} e & f \\ g & h \end{bmatrix} = [C_{21}] + [C_{22}] \tag{2}$$

Each matrix $[X_1]$ and $[X_2]$ is decomposed in accordance with the general relation for SVD of size 2×2 [6]:

$$[X] = \begin{bmatrix} x_{11} & x_{12} \\ x_{21} & x_{22} \end{bmatrix} = [C_1] + [C_2] = \sqrt{\frac{\omega + A}{8A^2}} \begin{bmatrix} P_1 & P_2 \\ P_3 & P_4 \end{bmatrix} + \sqrt{\frac{\omega - A}{8A^2}} \begin{bmatrix} P_4 & -P_3 \\ -P_2 & P_1 \end{bmatrix}, \tag{3}$$

where $x_{11}, x_{12}, x_{21}, x_{22}$ - pixels values (elements of the matrix $[X]$),

$$\omega = x_{11}^2 + x_{12}^2 + x_{21}^2 + x_{22}^2, \ A = \sqrt{v^2 + 4\eta^2}, \ v = x_{11}^2 + x_{21}^2 - x_{12}^2 - x_{22}^2,$$

$$\eta = x_{11}x_{12} + x_{21}x_{22}, \ \mu = x_{11}^2 + x_{12}^2 - x_{21}^2 - x_{22}^2,$$

$$P_1 = \sqrt{(A+\mu)(A+v)}, \ P_2 = \sqrt{(A+\mu)(A-v)}, \ P_3 = \sqrt{(A-\mu)(A+v)}, \ P_4 = \sqrt{(A-\mu)(A-v)}.$$

The matrices $[C_1]$ and $[C_2]$ from Eq. (3) depend on 4 parameters only (ω, v, μ, η), hence, the decomposition is of the kind "non over complete".

The so obtained matrices $[C_{i,j}]$ of size 2×2 for i, j = 1, 2 (as given in Eq. 2), are rearranged in new couples in correspondence with their energy. After the rearrangement, the first couple of matrices $[C_{11}]$ and $[C_{21}]$, which have high energy, defines the tensor $T_{1(2\times2\times2)}$, and the second couple $[C_{12}]$ and $[C_{22}]$ which have lower energy - the tensor $T_{2(2\times2\times2)}$.

$$T_{2\times2\times2} = T_{1(2\times2\times2)} + T_{2(2\times2\times2)} \tag{4}$$

After mode-2 unfolding both tensors in horizontal direction, is obtained:

$$\text{unfold } T_{1(2\times2\times2)} + \text{unfold } T_{2(2\times2\times2)} = \big[[X_{11}] \ [X_{21}] \big] + \big[[X_{12}] \ [X_{22}] \big] \tag{5}$$

In the second level of $HSVD_{2\times2\times2}$, on each matrix $[X_{i,j}]$ of size 2×2 is applied $SVD_{2\times2}$, and in result is obtained:

$$[X_{11}] = [C_{111}] + [C_{112}], \ [X_{21}] = [C_{211}] + [C_{212}], \ [X_{12}] = [C_{121}] + [C_{122}], \ [X_{22}] = [C_{221}] + [C_{222}]. \tag{6}$$

The so calculated matrices $[C_{i,j,k}]$ of size 2×2 for i, j, k = 1, 2 are rearranged into 4 new couples with similar energy in order, following the energy decrease. Each of these 4 couples of matrices defines a corresponding tensor of size $2 \times 2 \times 2$. After their unfolding is obtained:

$$\text{unfold } T_{1(2\times2\times2)}(1) + \text{unfold } T_{1(2\times2\times2)}(2) + \text{unfold } T_{2(2\times2\times2)}(1) + \text{unfold } T_{2(2\times2\times2)}(2)$$
$$= \big[[C_{111}] \ [C_{211}] \big] + \big[[C_{121}] \ [C_{221}] \big] + \big[[C_{112}] \ [C_{212}] \big] + \big[[C_{122}] \ [C_{222}] \big]. \tag{7}$$

In result of the execution of the two $HSVD_{2\times2\times2}$ levels, the tensor $T_{2\times2\times2}$ is represented as:

$$T_{2\times2\times2} = T_{1(2\times2\times2)}(1) + T_{1(2\times2\times2)}(2) + T_{2(2\times2\times2)}(1) + T_{2(2\times2\times2)}(2) = \sum_{i=1}^{2}\sum_{j=1}^{2} T_{i(2\times2\times2)}(j). \qquad (8)$$

2. Calculation of the Radix-(2×2) $HSVD_{4\times4\times4}$ for tensor, representing image of size $4 \times 4\times4$

In the first level of the $HSVD_{4\times4\times4}$, the tensor $T_{4\times4\times4}$ (for $N = 4$) from Fig. 3 is divided into eight elementary tensors (kernels) of size $T_{2\times2\times2}(i)$. Each elementary tensor is decomposed into 4 new tensors $T_{i(2\times2\times2)}(j)$, for $j = 1, 2, 3, 4$ of same size.

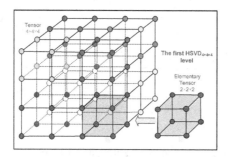

Fig. 3. The tensor $T_{4\times4\times4}$ is divided into eight elementary tensors $T_{2\times2\times2}(i)$ for $i = 1, 2, .., 8$. (Color figure online)

In the 1st level of $HSVD_{4\times4\times4}$, on each group of 8 pixels of same color (yellow, red, green, blue, white, purple, light blue, and orange) is applied $HSVD_{2\times2\times2}$. As a result, four tensors $T_{i(4\times4\times4)}(j)$ are obtained.

In the second level of the $HSVD_{4\times4\times4}$, each of these four tensors $T_{i(4\times4\times4)}(j)$ is divided into eight kernels $T_{i,k(2\times2\times2)}(j)$ for $i = 1, 2, j = 1, 2$ and $k = 1, 2$ in the way, defined by the spatial net for pixel interlacing as shown on Fig. 4. The color of the pixels in each kernel corresponds to that from the first level of the $HSVD_{4\times4\times4}$ algorithm. On each kernel is applied the $HSVD_{2\times2\times2}$ algorithm.

After the execution of the 1st decomposition level, the tensor $T_{4\times4\times4}$ is represented as a sum of 4 components:

$$T_{4\times4\times4} = \sum_{i=1}^{2}\sum_{j=1}^{2} T_{i(4\times4\times4)}(j) \qquad (9)$$

The so calculated 4 tensors $T_{i(4\times4\times4)}(j)$ are arranged in correspondence with the mean singular values (energy) of the kernels $T_{i,k(2\times2\times2)}(j)$, for $i = 1, 2, j = 1, 2, k = 1, 2$. The tensors $T_{i(4\times4\times4)}(j)$ are rearranged in accordance with the energy decrease of the kernels $T_{i,k(2\times2\times2)}(j)$, which build them. On each kernel is applied again the two-level $HSVD_{2\times2\times2}$ in accordance with Fig. 2.

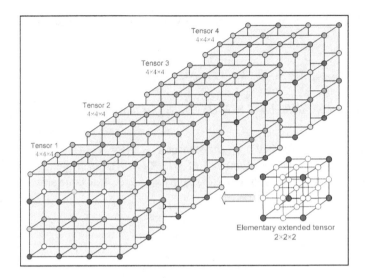

Fig. 4. Arrangement of tensors $T_{i(4\times4\times4)}(j)$ into kernels $T_{j(2\times2\times2)}(i)$ in the second $HSVD_{4\times4\times4}$ level, where the $HSVD_{2\times2\times2}$ is applied on each group of pixels of same color (32 in total) (Color figure online)

After the execution of the second decomposition level, the tensor $\mathbf{T}_{4\times4\times4}$ is represented as a sum of 16 components:

$$\mathbf{T}_{4\times4\times4} = \sum_{i=1}^{4} \sum_{j=1}^{4} \mathbf{T}_{i(4\times4\times4)}(j) \tag{10}$$

The computational graph of the 2-level $HSVD_{4\times4\times4}$ decomposition is shown on Fig. 5. The so obtained 16 tensors $\mathbf{T}_{i(4\times4\times4)}(j)$ are arranged, following the decreasing values of the singular values (energies) of the kernels, $\mathbf{T}_{i,k(2\times2\times2)}(j)$, which compose them, for $i = 1, 2, j = 1, 2$ and $k = 1, 2$. In the first level of the full $HSVD_{4\times4\times4}$, the $SVD_{2\times2}$ is executed 32 times, and in the second level - 64 times.

3. Calculation of the Radix-(2×2) HSVD for tensor, representing image of size $N \times N \times N$ for $N = 2^n$

The decomposition of the tensor $\mathbf{T}_{4\times4\times4}$ could be generalized for the case, when the tensor $\mathbf{T}_{N\times N\times N}$ is of size $N \times N \times N$ for $N = 2^n$. As a result of the use of the Radix-(2×2) $HSVD_{N\times N\times N}$ algorithm, the tensor $\mathbf{T}_{2^n\times2^n\times2^n}$ is represented as a sum of $N^2 = 2^n$ eigen tensors:

$$\mathbf{T}_{2^n\times2^n\times2^n} = \sum_{i=1}^{2^n} \sum_{j=1}^{2^n} \mathbf{T}_{i(2^n\times2^n\times2^n)}(j) \tag{11}$$

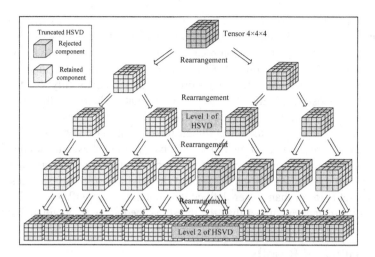

Fig. 5. Structure of the computational graphs of the full and truncated binary tree for execution of the 2-level algorithm for the $HSVD_{4\times4\times4}$ based on the $SVD_{2\times2}$

The eigen tensors $\mathbf{T}_{i(2^n\times2^n\times2^n)}(j)$ of size $2^n \times 2^n \times 2^n$ for i, j, k = 1, 2, .., 2^n are arranged in accordance with the decreasing values of the kernels energies $\mathbf{T}_{i,k(2\times2\times2)}(j)$, which build them. The number of hierarchical levels needed for the execution of the $HSVD_{N\times N\times N}$ algorithm (for $N = 2^n$), is $n = lg_2N$. The number of retained singular tensors is two times lower than that for Truncated $HSVD_{N\times N\times N}$, and in the last decomposition level it is $N^2/2 = 2^{2n-1}$.

6 Face Recognition

After successfully locating a face using the Viola & Jones method previously described, the faces must be analyzed to extract their features, which are then used to identify the person as either known or unknown. Much like face detection, there are many ways to perform facial recognition and each has its strengths and weaknesses. The system outlined in this paper uses an implementation of TensorFaces for facial recognition, as described in Sect. 5. The Tensorface approach has the advantage of being both fast and reasonably accurate, although it struggles with extreme variance in lighting and loses accuracy when objects are occluding the major identifiers (eyes and mouth). State of the art facial recognition methods, such as 3D facial analysis can overcome these issues, but they are computationally intensive and require expensive specialized equipment.

The TensorFace method requires a training set of faces to compare any new faces to. All of the images in the training set must be converted to grayscale and cropped to only contain the face of the subject; the images must also be the same size. Every image in the training set will be represented as an N × M matrix, where N is the length of an image in vector format and M is the total number of images in the set. Unfortunately, the dimensionality of the images is very high: for a small image that is only 200×200 pixels, there are 40,000 dimensions. A common technique for reducing the dimensionality of a dataset called Principal Component Analysis (PCA) is used to keep only the features that best describe the images. PCA is useful for working with images because it finds all of the eigenvectors, which are orthogonal, and keeps only the dimensions which hold the most information; while some information is lost in this process, the dimensionality of a dataset can be reduced drastically. Once PCA is applied and all of the images are represented in the lower dimensionality subspace, the system is ready to try to recognize faces. Any unknown faces will be represented as a linear combination of the TensorFace in the training set, with a vector of coefficients that correspond to the weights of each TensorFace. If the distance of the unknown face to its nearest face in the training set is lower than a user-defined threshold, the face will be positively identified but if the distance is higher than the threshold the face will remain unknown.

7   Experimental Results

To test the execution speed and accuracy of the proposed facial recognition surveillance system two different experiments were applied. In the first set of experiments YALE dataset and ORL dataset were used.

Tables 1 and 2 show the recognition rate for different number of training and testing images using Singular Values Decomposition with Hidden Markov Model (SVD-HMM) system [9].

Table 1.   Recognition rate for different number of training and testing images – ORL database

ORL database		
Number of images		Recognition rate
For training	For testing	(%)
205	205	96.6%
246	164	96.3%
287	123	96.7%
328	82	98.8%
369	41	97.6%

Table 2.   Recognition rate for different number of training and testing images – YALE database

YALE database		
Number of images		Recognition rate
For training	For testing	(%)
75	75	82.7%
90	60	80%
105	45	73.3%
120	30	90%
135	15	100%

While the ORL face recognition accuracy is practically not affected, the YALE dataset is more unstable in the presence of such variations.

Table 3 shows the recognition rate for different number of training and testing images using Radix-(2 × 2) HSVD, for YALE dataset recognition accuracy is completely affected. The table shows that the rate is increased and reached 100% along with changing of the number of training and testing images.

Table 3. Recognition rate using Radix-(2×2) Hierarchical Singular Value Decomposition

YALE Database		
Number of images		Recognition rate (%)
For training	For testing	
75	75	89.45%
90	60	97.9%
105	45	98.43%
120	30	100%
135	15	100%

The 100% recognition rate shown on Table 3 follows from the use of 3D HSVD instead of using 2D SVD.

Figure 6 shows the 38 subjects of YALE dataset. In the experiment 20 images for each subject are used. Figure 7 shows 15 TensorFaces basic vector from HSVD tensor decomposition using YALE dataset.

Fig. 6. The facial images of YALE database used for this experiment

In the second experiment multiple tests were performed using a short video clip of students entering a room. Although the system was designed to operate on a live feed, using a video clip is necessary to compare the execution time and accuracy of the different methods used in the program. A total of four experiments were run on the video clip, one for each combination of a face detection cascade (LBP or Haar Cascades) and motion detection (on or off). The initial results show that there is definite increase in speed when using the motion detection, but this increase in speed is mitigated when there is a large amount of noise or movement (such as a sudden change in camera focuses). Local binary patterns drastically reduce the execution time of the program

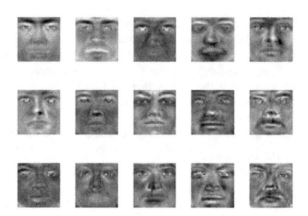

Fig. 7. Group of 15 TensorFaces basic vector from HSVD tensor decomposition using YALE dataset

compared to Haar cascades, but fail to detect the face as often. Both cascade types fail to detect profile-shots of the face as well as rotation of the head.

8 Conclusions

The results of using LBP or Haar cascades in a surveillance system show that they are not ideal. Even when the surveillance camera is positioned strategically to reduce the chances of rotated or non-frontal faces appearing in a frame, LBP and Haar cascades fail to detect faces far too often, especially when the subject is moving quickly or moving their head. Other methods, such as Speeded Up Robust Features (SURF) [10] which is rotation invariant is better suited to the task of facial detection and recognition. Use of motion detection to narrow the region of interest provided a noticeable reduction in execution time and could be combined with most other facial detection algorithms for the same effect; motion detection is almost a perfect fit for these surveillance scenarios since they often involve cameras that are at a fixed position with a mostly static background.

Unlike the Multilinear SVD, new Radix-(2 × 2) HSVD does not require iterative calculations. The computational complexity of the algorithm for the decomposition of the tensor $\mathbf{T}_{2^n \times 2^n \times 2^n}$, represented by the number of needed arithmetic operations, is $O(2^{4n})$, i.e., it is approximately 3 times lower than the number of operations $O(3 \times 2^{3n} + 3 \times 2^{4n})$, needed for the H-Tucker tensor decomposition [5], which represents the MSVD. However, the new decomposition needs larger memory (about 1/3) than the H-Tucker tensor decomposition.

According to Eqs. (3) and (8), in each hierarchical level the decomposition of the elementary tensors $2 \times 2 \times 2$ is repeatedly executed. This permits to implement the decomposition by using similar sets of calculations, executed in parallel. As a result, the use for the first time of the Radix-(2 × 2) HSVD in the face recognition systems, offers better abilities for real-time applications than the famous tensor decompositions.

Acknowledgment. This work was supported by the Research Experiences for Undergraduates (REU) Program of the National Science Foundation under Award Number 1359323.

References

1. Bebawy, M., Anwar, S., Milanova, M.: Active shape model versus deep learning for facial recognition in security. In: MPRSS-4-2016, Cancun, Mexico, December 4, 2016 (accepted)
2. Hu, G., Yang, Y., et al.: When face recognition meets with deep learning: an evaluation of convolution neural networks for face recognition. In: ICCV 2015, pp. 142–150 (2015)
3. Simplicam. A Home Surveillance Camera, Internet. http://www.simplicam.com/
4. Cichocki, A., Mandic, D., Phan, A., Caiafa, C., Zhou, G., Zhao, Q., De Lathauwer, L.: Tensor decompositions for signal processing applications. IEEE Signal Process. Mag. **32**(2), 145–163 (2015)
5. Kountchev, R., Kountcheva, R.: Radix-(2 × 2) hierarchical SVD for multi-dimensional images. In: Proceedings of the IEEE International Conference on Telecommunications in Modern Satellite, Cable and Broadcasting Services (TELSIKS 2015), Nis, Serbia, 14–17 October, 2015, pp. 45–55 (2015)
6. Kountchev, R., Kountcheva, R.: New approaches for hierarchical image decomposition, based on IDP, SVD, PCA and KPCA. In: Kountchev, R., Nakamatsu, K. (eds.) New Approaches in Intelligent Image Analysis. ISRL, vol. 108, pp. 1–58. Springer, Cham (2016). doi: 10.1007/978-3-319-32192-9_1
7. Lu, H., Plataniotis, K., Venetsanopoulos, A.: MPCA: multilinear principal component analysis of tensor objects. IEEE Trans. Neural Netw. **19**(1), 18–39 (2008)
8. Viola, P., Jones, M.: Rapid object detection using a boosted cascade of simple features. In: Computer Vision and Pattern Recognition (CVPR 2001), pp. 511–518 (2001)
9. Dinkova, P., Georgieva, P., Milanova, M.: Face recognition using singular value decomposition and hidden markov model. In: 16th WSEAS International Conference on Mathematical Methods, Computational Techniques and Intelligent Systems (MAMECTIS 2014), Lisbon, Portugal, October 30–November 1, 2014, pp. 144–149 (2014)
10. Bay, H., Tuytelaars, T., Gool, L.V.: Speeded-up robust features (SURF). Comput. Vis. Image Underst. **110**(3), 346–359 (2008)
11. Grasedyck, L.: Hierarchical singular value decomposition of tensors. SIAM J. Matrix Anal. Appl. **31**(4), 2029–2054 (2010)
12. Yang, M.-H., Kriegman, D.J., Ahuja, N.: Detecting faces in images: a survey. PAMI **24**(1), 34–58 (2002)
13. OpenCV. http://opencv.org/

Performance Analysis of Gesture Recognition Classifiers for Building a Human Robot Interface

Tiziana D'Orazio, Nicola Mosca, Roberto Marani, and Grazia Cicirelli[✉]

Institute of Intelligent Systems for Automation (ISSIA),
National Research Council of Italy, Bari, Italy
{dorazio,cicirelli}@ba.issia.cnr.it

Abstract. In this paper we present a natural human computer interface based on gesture recognition. The principal aim is to study how different personalized gestures, defined by users, can be represented in terms of features and can be modelled by classification approaches in order to obtain the best performances in gesture recognition. Ten different gestures involving the movement of the left arm are performed by different users. Different classification methodologies (SVM, HMM, NN, and DTW) are compared and their performances and limitations are discussed. An ensemble of classifiers is proposed to produce more favorable results compared to those of a single classifier system. The problems concerning different lengths of gesture executions, variability in their representations, generalization ability of the classifiers have been analyzed and a valuable insight in possible recommendation is provided.

Keywords: Gesture recognition · Feature extraction · Model learning · Gesture segmentation · Ensemble classifier · Human-robot interface

1 Introduction

In the last decades, gesture recognition has been attracting a lot of attention as a natural way to interact with computer through intentional movements of hands, arms, face, or body. A number of approaches have been proposed giving particular emphasis on hand gestures and facial expressions by the analysis of images acquired by conventional RGB cameras [1,2].

The recent introduction of low cost depth sensors (ToF cameras), allowed the spreading of new gesture recognition approaches and the possibility of developing personalized human computer interfaces. Depth images provide the 3D structure of the scene which can be easily used to simplify many tasks such as people segmentation and tracking, body part recognition, motion estimation and so on. Recent reviews on human activity recognition and motion analysis from 3D data have been published in [3–6]. At present, Gesture Recognition through visual and depth information is one of the main active research topics in the computer vision community. The Kinect provides synchronized depth and color (RGB) images where each pixel corresponds to an estimate of the distance between

© Springer International Publishing AG 2017
F. Schwenker and S. Scherer (Eds.): MPRSS 2016, LNAI 10183, pp. 60–72, 2017.
DOI: 10.1007/978-3-319-59259-6_6

the sensor and the closest object in the scene together with the RGB values at each pixel location. Together with the sensors some software platforms are available to detect and track one or more people in the scene and extract the corresponding human skeleton in real time. The availability of information about joint coordinates and orientation has provided a great impulse to research [7–12].

Many papers, presented in literature in the last years, use normalized coordinates of proper subset of skeleton joints which are able to characterize the movements of the body parts involved in the gestures [13,14]. Angular information between joint vectors have been used to maximize the invariance of the skeletal representation with respect to the camera position [15].

Different methods have been used to generate gesture models. Hidden Markov Models (HMM) are a common choice for gesture recognition as they are able to model sequential data over time [16–18]. Usually HMMs require sequences of discrete symbols, so different quantization schemes are first used to quantize the features which characterize the gestures. In [16] a K-means clustering is used to convert the feature vectors (joint angles) into the observable symbols for HMMs. In [17] a uniform quantization of the orientations in 12 sectors (every $\pi/6$) has been used. In [18] the skeleton coordinates are transformed into feature sequences by considering the features as observations of Gaussian distributions. Support Vector Machines (SVM) reduce the classification problem into multiple binary classifications either by applying a one-versus-all (OVA-SVM) strategy (for a total of N classifiers for N classes) [19] or a one-versus-one (OVO-SVM) strategy (for a total of $N \times (N - 1)/2$ classifiers for N classes) [20,21]. Artificial Neural Networks (NNs) represent another alternative methodology to solve classification problems in the context of gesture recognition [22]. The choice of the network topology, the number of nodes/layers and the node activation functions depends on the problem complexity and can be fixed by using iterative processes which run until the optimal parameters are found [23]. Distance-based approaches, starting from the assumption that the features characterizing the models are well separated, apply distance metrics to measure the similarity between samples and gesture models. These methods have to solve the problem of variable lengths of the sequences of features in order to apply any metric for comparisons. Several solutions have been proposed either transforming the length of features in common reference space (such as Dynamic Time Warping techniques (DTW) [24]), or using ad hoc procedure to align the sequences [25].

In this paper we focus on the development of a Gesture Recognition approach for developing a natural human computer interface. The final aim is to remotely control a mobile robot by recognizing gestures performed by the users. Each gesture defines a particular command for the robot. The low cost Kinect camera is used to acquire the 3D information of the scene, then the features which describes the gesture are extracted and are fed into the classification module. Once the gesture is recognized, its associated command is sent to the robot.

In order to obtain the best performance, a study of different classification approaches has been carried out. In this paper we compare different classification methods such as Dynamic Time Warping (DTW), Neural Network (NN),

Support Vector Machine (SVM) and Hidden Markov Model (HMM). The performance of each methodology has been evaluated considering several users making the gestures. This performance analysis is required as different users perform gestures in a personalized way and with different velocity, furthermore even the same user executes gestures differently in separate sessions. So, in order to build an efficient and robust human robot interface the classifier must have good generalization ability. In conclusion, the main contribution of this paper is twofold: first the analysis of the performance of different classifiers is given, then a combination of classifiers is applied in order to increase gesture recognition accuracy.

The rest of the paper is organized as follows. The overall description of the problem and the definition of the gestures are given in Sect. 2. The definition of the features is provided in Sect. 3. The methodologies selected for the gesture model generation are describe in Sect. 4. Finally Sect. 5 presents the experiments carried out. Section 6 reports some conclusions.

2   Definition of the Problem and Target Gestures

In this paper we have considered all the problems related to the development of a gesture recognition interface which can be used in a real context and with low cost sensors. The Kinect camera with OpenNI Libraries is used to extract people in the environment and to segment the body parts involved in the movement. Ten gestures have been defined and have been performed in front of the camera by using the left arm. Figure 1 shows the gestures that have been chosen for the experiments. Throughout the paper we will refer to these gestures by using the following symbols G_1, G_2, G_3, ... G_N, where $N = 10$. Some of these gestures involve movements in a plane parallel to the camera (G_1, G_3, G_4, G_7) while the others involve a forward motion in a plane perpendicular to the camera (G_2, G_5, G_6, G_8, G_9, G_{10}). Furthermore, some gestures are quite similar in terms of variations of joint coordinates, the only difference is the plane in which they are performed (see for example G_9 and G_4, G_1 and G_8).

Another point to consider is the complexity of gestures in terms of feature variations. The human skeleton extracted by the Kinect framework is quite unstable when the user has not a well distinguishable shape. For this reason, those gestures involving a forward motion have strong variations that mainly depend on the instability of the joint positions.

The development of a natural interface for gesture recognition involves two main challenging tasks: the spatial and temporal resolution of gestures and the generalization ability of the gesture classifier. In this paper these problems have been investigated and tackled. First the best features, which are more representative and discriminant for the chosen set of gestures, are selected. Then, an algorithm based on Fast Fourier Transform (FFT) has been applied to estimate the duration of each gesture execution. Finally, different methodologies have been compared for the evaluation of gesture model generalization. Since each classifier works in a different way for all the gestures, a combination of classifiers is proposed to reduce the risk of wrong decision.

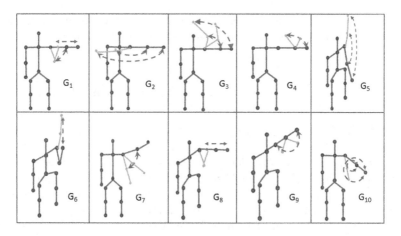

Fig. 1. Ten different gestures are shown. Gestures G_1, G_3, G_4 and G_7 involve movements in a plane parallel to the camera. Gestures G_2, G_5, G_6, G_8, G_9, and G_{10} involve a forward motion in a plane perpendicular to the camera.

3   Feature Selection

The complexity of the gestures strictly affect the feature selection and the methodology for the gesture model generation. If the gestures are distinct enough, the recognition can be easy and reliable. So, joint coordinates which are immediately available by the Kinect software platforms, could suffice. Only a normalization is required to guarantee invariance with respect to the user's height, arm length, distance and orientation. On the other hand, the angular information of joint vectors have the great advantage of maximizing the invariance of the skeletal representation with respect to the camera position. In [26] the angles between the vectors generated by the elbow-wrist joints, and the shoulder-elbow joints, are used to generate the models of the gestures. However, the experiments prove that these features are not discriminant enough to separate all the gestures. For this reason, in our approach, we use features which are more complex, but more representative since the orientation of a rigid body in a 3-dimensional space is considered. So, we use the quaternions of two arm joints (returned by the OpenNi framework), in particular of the shoulder and elbow left joints. A quaternion is a set of numbers that comprises a four-dimensional vector space and is denoted by:

$$q = a + bi + cj + dk$$

where the coefficients a, b, c, d are real numbers and i, j, k are imaginary units. The quaternion q represents an easy way to code any 3D rotation expressed as a combination of a rotation angle and a rotation axis. The quaternions of the shoulder and elbow left nodes produce a feature vector for each frame i defined by:

$$V_i = [a_i^s, b_i^s, c_i^s, d_i^s, a_i^e, b_i^e, c_i^e, d_i^e]$$

where the index s stands for shoulder and e stands for elbow. The sequence of vectors which characterizes the whole gesture is given by the vector

$$\overline{V} = [V_1, V_2, ..., V_n]$$

where n is the number of frames during which the gesture is entirely performed.

4   Gesture Classification

In this paper different methodologies for the gesture classification have been applied. For each of them, the best parameter configuration and the best architecture topology which assure the convergence of each method have been selected. Neural Networks (NNs), Support Vector Machines (SVMs), Hidden Markov Models (HMMs) and Dynamic Time Warping (DTW) have been recognized as those most promising for gesture recognition.

Some of these methodologies require a fixed length for the input vector. For this reason, as the length of a gesture can vary either if the gesture is executed by the same user or by different users, a preliminary step of gesture length normalization is required. A Fast Fourier Transform (FFT)-based approach has been applied. Repeating a gesture for a number of times, it is possible to approximate the sequence of features as a periodic signal. So users are asked to repeat the same gesture without interruption and all the frames of the sequences are recorded. Applying the (FFT) and by tacking the position of the fundamental harmonic component, the period can be evaluated as the reciprocal value of the peak position. The estimated period is then used to interpolate the sequence of feature vectors in a fixed number of values which can be provided as input to the classifiers. Several sequences of gestures, executed by different people, have been acquired in order to construct training and testing sets. Furthermore during the real application of the human robot interface, as it is not possible to know the starting frame and the length of the gestures executed by the user, FFT is also applied to estimate the period p as previously described. So applying a sliding window approach on the video sequence, segments of p frames are extracted and resized with linear interpolation in order to have a number of frames equal to the length of the gesture used in the training phase. Then the features are extracted and fed into all the classifiers.

4.1   Neural Network

Ten different Neural Networks have been used to generate the models of the ten considered gestures. Each ANN architecture has an input layer, one 40-nodes hidden layer and a single node in the output layer. The back-propagation algorithm is applied for model learning. Ten different training sets are constructed. Each training set contains the feature vectors of one gesture as positive examples and those of all the other gestures as negative ones. As each gesture execution lasts a different number of frames, a preliminary normalization of the sequence length has been applied. The length of each gesture as been fixed to $n = 60$

frames, where n has been evaluated considering the average length of all gestures (about 2 s). So the input vector for the ANNs has been defined by the elements of feature vector Vi times the number of frames n. After training, a test phase has been carried out. Each gesture sample is provided to all the 10 ANNs and is assigned to the winning class, i.e. the one with the maximum output among all the ANNs.

4.2 Support Vector Machine

In the case of SVM, a one-versus-one method has been applied. This method builds one binary classifier for every pair of gestures, so a total of $N(N-1)/2$ two-class classifiers (45 classifiers in our case). During the training phase, the classifiers learn the optimal hyperplane which separates the two classes. The testing phase is carried out by a max-wins voting strategy in which every classifier assigns the test gesture to one of the two classes, then the vote of the win class is increased by one. After each of the 45 classifiers makes its vote, the max-wins voting strategy assigns the sample to the class with the largest number of votes. As in the case of ANNs, the input vectors for SVMs are the normalized n-length vectors of Vi features.

4.3 Hidden Markov Models

Ten different discrete Hidden Markov Models have been built to generate the models of the defined gestures. The choice of the number of hidden states depends on the complexity of the process that has to be modelled. Different numbers of states have been used and the best of these has been selected. A fully connected HMM topology and the Baum-Welch algorithm have been applied to learn the optimal model parameter. For both training and testing samples, a K-means approach has been used to associate each continuous values of the feature vectors with a discrete value. The preliminary normalization step has been applied to gesture samples, too. As in the case of ANNs, during testing each gesture sample is inputted to the 10 trained HMMs and the HMM showing the maximum probability of the data is selected as the winning class.

4.4 Dynamic Time Warping

DTW algorithm allows to compute the distance between two signals in terms of their associated feature values. The Euclidean Distance has been used to find the optimal alignment between the different sequences of feature vectors and solve the classification problem. In this case, the normalization of feature vectors is not necessary due to the *warping* peculiarity of DTW. A training phase is also not needed apart from the selection for each gesture of a target feature sequence which is representative of each considered gesture. The target gestures have been selected applying the DTW to the set of training samples inside each gesture class. The one with the minimum distance from all the other samples of the same class has been chosen as target gesture. During testing phase, instead, DTW is used to compare each test instance of gesture with all the target ones computing the relative distances. The winning class is that of the closest target gesture.

4.5 Diversity of Classifiers

In the ideal case of noiseless input data, all the previous classifiers have perfect generalization performances. Such a generalization is quite impossible for several reasons: noise in data acquisition phase; variability of the gesture templates both if performed by different users or by the same one (even the same user does not perform gestures exactly the same each time). At the best, different classifiers produce good results most of the time, and due to their different nature, they behave differently on different instances of gestures. In this case, a combination of classifiers can reduce the total error. A measure of diversity among classifiers is necessary in order to assess their diversity and justify their combination. Several measures have been defined in literature [28], the simplest one is the pair-wise diversity measure between two classifiers. Given two classifiers C_i and C_j we indicate with a the fraction of instances correctly classified by both classifiers; with b the fraction of instances correctly classified by C_i but incorrectly by C_j; with c the fraction of instances incorrectly classified by the C_i but correctly by C_j, and with d the fraction of instances incorrectly classified by both classifiers. The Q statistic [27] measures the diversity between the two classifiers as following:

$$Q_{i,j} = \frac{ad - bc}{ad + bc}$$

Q values close to zero indicate maximum diversity of classifiers, positive Q values indicate that both classifiers correctly classifies the same instances, whereas negative Q values indicate that different errors are done on different instances by the classifiers.

4.6 Combining Classifiers

A strategic combination of classifiers can reduce the total errors, improving the performance of a single classifier. Let us define the decision of the tth classifiers as $d_{t,j} \in \{0,1\}$ for $t = 1,..,T$ (T is the number of classifiers) and $j = 1,...,N$ (N is the number of classes). Value 1 indicates that classifier t chooses class j, whereas value 0 indicates all the other cases. The majority voting scheme [28] chooses the class which receives the highest number of votes. In many cases, as the vote cannot be distributed in an unequivocal way, for example when the same vote is obtained for more classes, the final decision cannot be made. For this reason, a weighted majority voting scheme has been used weighting more heavily the more expert classifiers in order to improve the overall performance. So the class of the gesture G_J is chosen if

$$\sum_{t=1}^{T} w_t d_{t,J} = max_{j=1}^{N} \sum_{t=1}^{T} w_t d_{t,j} \tag{1}$$

where w_t is the weight of classifier t which measures the quality of the classifier decision.

5   Experiments and Evaluations

Different experiments have been carried out in order to evaluate the performance of the proposed framework.

In order to manage real-world situations, several sequence of gestures performed by different users in front of a Kinect camera have been acquired. The use of public datasets has been discarded as they do not assure that real situations are handled, furthermore they contain few executions of the sample gestures which are mainly acquired in the same conditions.

User 1										
%	G1	G2	G3	G4	G5	G6	G7	G8	G9	G10
DTW	100	96,88	100	100	100	100	73,68	100	94,74	80
NN	100	100	100	100	88,89	100	100	100	100	100
SVM	100	100	100	100	100	100	100	100	100	100
HMM	100	100	11,43	100	72,22	100	84,21	100	100	60

User 2										
%	G1	G2	G3	G4	G5	G6	G7	G8	G9	G10
DTW	100	47,06	100	100	63,16	100	65	100	100	0
NN	100	94,12	100	96,15	94,74	100	65	90,48	90,48	100
SVM	100	94,12	100	100	100	100	65	100	100	100
HMM	100	88,24	100	100	78,95	100	65	85,71	95,24	90

User 3										
%	G1	G2	G3	G4	G5	G6	G7	G8	G9	G10
DTW	53,85	100	96,67	100	93,94	100	93,94	100	78,79	72,22
NN	100	65,38	100	100	100	100	100	97,06	3,03	100
SVM	100	73,08	60	100	100	100	100	94,12	9,09	91,67
HMM	100	88,24	100	100	78,95	100	65	85,71	95,24	90

User 4										
%	G1	G2	G3	G4	G5	G6	G7	G8	G9	G10
DTW	96,55	100	57,14	100	94,44	100	90,91	100	36,36	52,38
NN	100	100	100	100	94,44	100	100	100	100	100
SVM	100	100	61,90	100	100	100	100	100	100	100
HMM	100	94,12	0	90,48	61,11	100	100	100	81,82	0

Fig. 2. Results of tests performed on gestures executed by four different Users: User1, User2, User3, and User4. In each table the rate of correctly recognized gestures, for each type of gesture and each classifier, are listed.

Gestures performed by one user (User 1) are used to build sets of samples for training the classifiers. For each gesture, a number of sequences containing several executions of the same gesture were recorded. As introduced in Sect. 4 a FFT-based algorithm has been applied to process each acquired sequence and evaluate the period in order to extract sub-sequences of frames, each containing one gesture execution. As previously stated a normalization is applied to each sub-sequence in order to obtain feature vectors with the same length useful for training ANNs, SVMs and HMMs. In the DTW case, all the training samples of each gesture class are used to select the target gesture as described in Sect. 4.4.

After the training phase, other sessions of acquisitions were carried out considering both User 1 and other three different users (referred as User 2, User 3 and User 4). The same processing applied to the training sets is applied to these new sequences in order to build different testing sets. It is important to observe that the execution of the gestures by different people in different sessions involve a number of factors that do not guarantee a uniformity of gesture execution with respect the training ones. These are: different relative positions between users and camera, different orientation of the arms, different amplitude of the movement, and so on. All these factors can greatly modify the resulting skeletons and the joint positions producing large variations in the extracted features.

In Fig. 2 the recognition rates, obtained during testing by applying the different classifiers, are reported for each user. As can be observed, there is a great variability of recognition rates: the classifiers answers differently for each type of gesture and for each user. This is principally due to the individual complexity of the gestures and to the subjective execution of the gestures by each user. As a consequence the choice of the best classifier (best in terms of the better classification rate) among the analyzed ones is not possible. So another experiment has been carried out in order to consider the possibility of improving the recognition performance and building a human computer interface as general as possible. To this aim the ensemble based algorithm has been applied for taking the final decision as described in Sect. 4.6. First a measure of diversity has been evaluated for each pair of classifiers as described in Sect. 4.5. In Table 1 the values of the Q statistics evaluated considering test results of User 1 are listed. Values close to 0 implies large diversity of classifiers, whereas in case of $Q = 0$ maximum diversity is obtained. This is the case, for example, of couples DTW-SVM, NN-SVM and SVM-HMM.

The Q values in Table 1 asses the diversity and uncorrelation of the classifiers making possible the application of an ensemble system. The idea is to combine the outputs of different uncorrelated classifiers in order to improve performance upon that of a single classifier. Therefore a new experiment has been carried out by applying the ensemble of classifiers approach which has been implemented by using the weighted majority voting scheme described in Eq. 1. The weights of each classifier for each type of gesture have been estimated considering the performance of the classifiers on the training set. Figure 3 shows the obtained performance rates after applying the ensemble based scheme. As can be seen in Fig. 3, the performance rates increase in all the cases. Weighting the decisions improve the overall performance as expected.

Table 1. Measures of the Q-values for each pair of classifiers.

	NN	SVM	HMM
DTW	−0.018	0	0.082
NN	-	0	0.17
SVM	-	-	0

Ensemble Decision										
%	G1	G2	G3	G4	G5	G6	G7	G8	G9	G10
User1	100	100	100	100	100	100	100	100	100	100
User2	100	100	100	100	84,62	100	100	100	100	100
User3	100	96,15	100	100	100	100	100	100	80,30	100
User4	100	100	76,19	100	100	100	100	100	100	100

Fig. 3. Results obtained applying the ensemble based algorithm. On average, the detection rates increase notably with respect to single classifiers.

Some important conclusions can be drawn considering the overall experiments carried out in this work: first, the solution of using only one user to train the classifiers can be pursued as the performance of classifiers on different users are quite good even if the gestures are performed in different ways. Second, the use of an ensemble based algorithm increases the performance of each individual classifier solving the unlucky situation in which one classifier completely fails. The last point concerns the evaluation of gesture complexity. Experiments prove that the failures are due to either the strict similarity between different gestures or to the complexity of the gestures. For example, gestures that involve a movement perpendicular to the camera image plane (such as G_2, G_5, G_6, G_8, G_9 and G_{10}) can produce false skeleton postures and consequently the extracted features are completely erroneous. For example, during the experiments, gesture G_8 is sometimes misclassified as gesture G_1. By observing Fig. 1, G_1 and G_8 actually involve the same movement, but in different planes: G_1 in the lateral plane, G_8 in the frontal one with respect to the camera. It is evident that a slight different orientation of the user in front of the camera could potentially produce miss-classifiable features.

Considering the results discussed above, a subset of gestures has been selected for building the human robot interface. Five of the ten considered gestures have been chosen for controlling the mobile platform. Table 2 lists these gestures and the associated commands for the robot. Figure 4 shows the scheme of the interface. As described in Sect. 4 an initialization stage is required in order to apply a FFT-based algorithm to estimate the length of gestures executed by the user. As soon as a gesture is recognized, a socket containing the code of the corresponding command is sent to the mobile robot controller to execute the action. A display informs the user about the correct reception of the command and plots the map of the environment with the current robot position to allow the user to change the command as soon as the previous task has been completed.

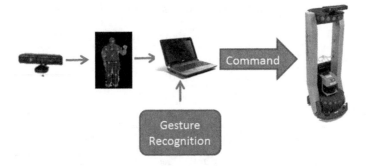

Fig. 4. Scheme of human robot interface.

Table 2. The gestures and the associated commands

Gesture	Command
G_1	Initialization
G_2	Home
G_4	Go to the Goal
G_6	Turn Around
G_7	Go Wondering
G_{10}	Stop

6   Discussion and Conclusions

In this paper we present a human computer interface based on a gesture recognition system by using the Kinect sensor. Quaternion features of the left shoulder and elbow joint are used to describe gestures. Different classification methods (DTW, NN, SVM, HMM) are used to construct the models of 10 different types of gestures executed by several users. In particular only the gestures executed by one user are used for building the training set, whereas all the others are used for the testing sets.

Several experiments have been carried out to compare the performance of each single classifier and to prove the great improvement obtained by applying an ensemble based decision method. This proves that the decision of using only one user for building the training set was not only convenient, but also successful. Furthermore the application of an ensemble based algorithm for making the final decision about the recognition of the gesture is very favorable as it solves problems related to erroneous recognition by the single classifiers. This is important if an efficient and robust human robot interface must be developed. Actually the conducted performance analysis of gesture recognition classifiers allows us to build an efficient human robot interface which uses the gestures for remotely controlling a mobile platform. Furthermore the proposed interface manages the problem of gesture length estimation. Actually, during online executions of gestures the starting frame and the length of the gestures are not *a-priori* known as they vary not only among the different types of gestures, but also among the different users. So we propose to use a FFT-based algorithm to solve this problem.

References

1. Murthy, G.R.S., Jadon, R.S.: A review of vision based hand gesture recognition. Int. J. Inform. Technol. Knowl. Manage. **2**(2), 405–419 (2009)
2. Hassan, M.H., Mishra, P.K.: Hand Gesture Modeling and ecognition using geometric features: a review. Canadian J. Image Process. Comput. Vis. **3**(1), 12–26 (2012)
3. Mitra, S., Acharya, T.: Gesture recognition: a survey. IEEE Trans. Syst. Man Cybern. Part C Appl. Rev. **37**(3), 311–324 (2007)
4. D'Orazio, T., Marani, R., Renó, V., Cicirelli, G.: Recent trends in gesture recognition: how depth data has improved classical approaches. Image Vis. Comput. **52**, 56–72 (2016)
5. Aggarwal, J.K., Xia, L.: Human activity recognition from 3D data: a review. Pattern Recogn. Lett. **48**, 70–80 (2014)
6. Chen, L., Wei, H., Ferryman, J.: A survey of human motion analysis using depth imagery. Pattern Recogn. Lett. **34**(15), 1995–2006 (2013)
7. Cruz, L., Lucio, F., Velho, L.: Kinect and RGBD images: challenges and applications. In: XXV SIBGRAPI IEEE Confernce and Graphics, Patterns and Image Tutorials, pp. 36–49 (2012)
8. Cheng, L., Sun, Q., Cong, Y., Zhao, S.: Design and implementation of human-robot interactive demonstration system based on kinect. In: 24th Chinese Control and Decision Conference (CCDC), pp. 971–975 (2012)
9. Almetwally, I., Mallem, M.: Real-time tele-operation and tele-walking of humanoid Robot Nao using Kinect Depth Camera. In: 10th IEEE International Conference on Networking, Sensing and Control (ICNSC), pp. 463–466 (2013)
10. Hachaj, T., Ogiela, M.R.: Rule-based approach to recognizing human body poses and gestures in real time. Multimedia Syst. **20**, 81–99 (2013)
11. Jacob, M.G., Wachs, J.P.: Context-based hand gesture recognition for the operating room. Pattern Recogn. Lett. **36**(15), 196–203 (2014)
12. Bodiroža, S., Doisy, G., Hafner, V.V.: Position-invariant, real-time gesture recognition based on Dynamic Time Warping. In: 8th ACM/IEEE International Conference on Human-robot Interaction, pp. 87–88 (2013)
13. Lai, K., Konrad, J., Ishwar, P.: A gesture-driven computer interface using Kinect. In: IEEE Southwest Symposium on Image Analysis and Interpretation (SSIAI) (2012)
14. Oh, J., Kim, T., Hong, H.: Using binary decision tree and multiclass SVM for human gesture recognition. In: IEEE International Conference on Information Science and Applications (ICISA) (2013)
15. Mangera, R.: Static gesture recognition using features extracted from skeletal data. In: 24th Annual Symposium of the Pattern Recognition Association of South Africa (2013)
16. Gu, Y., Do, H., Ou, Y., Sheng, W.: Human gesture recognition through a Kinect sensor. In: IEEE International Conference on Robotics and Biomimetics (ROBIO), pp. 1379–1384 (2012)

17. Wang, Y., Yang, C., Wu, X., Xu, S., Li, H.: Kinect based dynamic hand gesture recognition algorithm research. In: 4th International Conference on Intelligent Human-Machine Systems and Cybernetics (IHMSC), pp. 274–279 (2012)

18. Song, Y., Gu, Y., Wang, P., Liu, Y., Li, A.: A Kinect based gesture recognition algorithm using GMM and HMM. In: 6th International Conference on Biomedical Engineering and Informatics (BMEI), pp. 750–754 (2013)

19. Miranda, L., Vieira, T., Martinez, D., Lewiner, T., Vieira, A., Campos, M.: Online gesture recognition from pose kernel learning and decision forests. Pattern Recogn. Lett. **39**, 65–73 (2014)

20. Bhattacharya, S., Czejdo, B., Perez, N.: Gesture classification with machine learning using Kinect sensor data. In: 3rd International Conference on Emerging Applications of Information Technology (EAIT), pp. 348–351 (2012)

21. Althloothi, S., Mahoor, M.H., Zhang, X., Voyles, R.M.: Human activity recognition using multi-features and multiple kernel learning. Pattern Recogn. **47**, 1800–1812 (2014)

22. Ibraheem, N.A., Khan, R.Z.: Vision based gesture recognition using neural networks approaches: a review. Int. J. Hum. Comput. Interac. (IJHCI) **3**(1), 1–14 (2012)

23. Cicirelli, G., Attolico, C., Guaragnella, C., D'Orazio, T.: A kinect-based gesture recognition approach for a natural human robot interface. Int. J. Adv. Robot. Syst. **12** (2015)

24. Celebi, S., Aydin, A.S., Temiz, T.T., Arici, T.: Gesture recognition using skeleton data with weighted dynamic time warping. Comput. Vis. Theory Appl. **1**, 620–625 (2013)

25. Ding, I.J., Chang, C.W.: An eigenspace-based method with a user adaptation scheme for human gesture recognition by using Kinect 3D data. Appl. Math. Model. **39**(19), 5769–5777 (2015)

26. D'Orazio, T., Attolico, C., Cicirelli, G., Guaragnella, C.: A neural network approach for human gesture recognition with a kinect sensor. In: International Conference on Pattern Recognition Applications and Methods (ICPRAM), pp. 741–746 (2014)

27. Yule, G.: On the association of attributes in statistics. Phil. Trans. **194**, 257–319 (1990)

28. Polikar, R.: Ensemble based systems in decision making. IEEE Circuits Syst. Mag. **6**(3), 21–45 (2006)

On Automatic Question Answering Using Efficient Primal-Dual Models

Yusuf Osmanlıoğlu[1]([✉]) and Ali Shokoufandeh[2]

[1] Section of Biomedical Image Analysis,
University of Pennsylvania, Philadelphia, USA
yusuf.osmanlioglu@uphs.upenn.edu
[2] Department of Computer Science, Drexel University, Philadelphia, USA
ashokouf@cs.drexel.edu

Abstract. Automatic question answering has been a major problem in natural language processing since the early days of research in the field. Given a large dataset of question-answer pairs, the problem can be tackled using text matching in two steps: find a set of similar questions to a given query from the dataset and then provide an answer to the query by evaluating the answers stored in the dataset for those questions. In this paper, we treat the text matching problem as an instance of the inexact graph matching problem and propose an efficient approximate matching scheme. We utilize the well known quadratic optimization problem *metric labeling* as the framework of graph matching. In order to solve the text matching, we first embed the sentences given in natural language into a weighted directed graph. Next, we present a primal-dual approximation algorithm for the linear programming relaxation of the metric labeling problem to match text graphs. We demonstrate the utility of our approach on a question answering task over a large dataset which involves matching of questions as well as plain text.

Keywords: Metric labeling · Graph matching · Primal-dual approximation · Question answering

1 Graph Based Question Answering

Automated understanding of natural language is a problem studied under several disciplines including computer science, linguistics, and statistics. Turing initiated the research in automated natural language processing (NLP) with his seminal paper where he questioned whether computers can interpret human queries made in natural language and reply in a way indistinguishable from human response [28]. During the following decades, research in the field was mainly focused on complex rule based systems. The scalability of the rule sets has been the main limiting factor for the progress in this course. Research in the field flourished in late eighties with the advent of machine learning techniques such as neural networks and support vector machines (SVM) [16].

© Springer International Publishing AG 2017
F. Schwenker and S. Scherer (Eds.): MPRSS 2016, LNAI 10183, pp. 73–84, 2017.
DOI: 10.1007/978-3-319-59259-6_7

Question answering (QA) has been a major problem in NLP since the early days of research in the field [8]. From an information retrieval (IR) perspective, the main goal is to extract an answer from a large data source that satisfies a certain query [14]. It differs from other IR tasks in that queries being in natural language provide extra information due to syntactic and semantic relations between the words in contrast to simple lists of conventional search terms. This extra information can be utilized to represent sentences by directed graphs where words are expressed as nodes and relations among the words are illustrated by directed edges. Once the sentences are represented using graphs, graph matching methods can be utilized to tackle the QA problem.

In this paper, we approach question answering as an instance of the graph matching problem. More specifically, given a database of question-answer pairs and a query, the goal is to find a proper answer. To achieve this, we first embed the questions into graphs. Next, we find similar questions to the query by comparing the query graph to the graphs in the dataset. Finally, we determine the answer according to the answers of the similar questions from the dataset. We utilize the metric labeling approach for matching question graphs which we previously presented in [22] with its application to matching lineage trees in immune repertoires.

The rest of the paper is organized as follows: First, we provide an overview of the literature in natural language processing and question answering. Then, we introduce an embedding approach for representing sentences by directed weighted graphs. Next, we present the primal dual algorithm for metric labeling that is adapted to text matching. Finally, we provide an experimental evaluation of the proposed method over a question answering dataset and conclude with discussions and future work.

2   Background

The most basic method used for information retrieval and NLP tasks in general and QA in specific is the bag-of-words model [3]. In this model, individual words in a given text are statistically indexed while any relation that is present among the words is ignored. An extension to the bag-of words model that partially captures the syntactic relation between words is the n-grams, which represents the text as vectors where features are not only single words but chunks of words of size n [2]. Methods utilizing further syntactic information such as parse trees and part of speech tags are also devised for retrieval tasks [19]. Although successfully applied in various types of question answering, one major drawback of these methods is that they do not capture the semantic relations between words, which leads to challenges such as negation detection and coreference resolution [4,10]. In order to address these problems, machine learning methods such as recursive neural networks are commonly applied to NLP tasks in addition to various combinations of the approaches mentioned above [11,26].

Graph based methods are widely used in information retrieval in the NLP domain to overcome challenges that arise from ignoring the semantic relations between sentences and words by capturing syntactic and semantic relations together. Sentences can be embedded into directed graphs based on their structural properties with words represented as nodes and the relations between words represented as directed edges. Moschitti [20] only used dependency parse tree for defining relations among words. He used subtrees and subset trees of the parse tree as features for describing a sentence. Using these features, he trained an SVM for a predicate argument classification task. Utilizing coreference edges in addition to dependency parsing trees, graph matching techniques are applied to several NLP tasks including review relevance identification [25], inference recognition [21], and extraction of automatic reasoning chains [27].

Among several graph based methods applied to NLP tasks, we specifically mention two that utilize combinatorial optimization due to their relevance to our approach on QA. The first is due to Haghigi et al. [9] where the textual inference problem is formulated as a graph matching problem. In this study, sentences are first embedded into directed graphs by representing words as nodes and dependency relations between words as edges. Then, the matching is defined as a minimization problem with a two term objective function which consists of the cost of assigning vertices and the cost of substituting relations. This approach can be contrasted to the metric labeling problem in that it considers pairwise relations among vertices while establishing the mapping. The difference lies in the way that the effect of pairwise relations are incorporated into the matching. The second study is due to Pang and Lee [23] which utilizes an approach based on the metric labeling formulation for the rating-inference problem. Here, the goal is to decide the sentiment value of a given review with respect to a multi-point scale, such as one to five stars. In this comparative study, three methods are proposed for solving the multi-category classification of query sentences: one-vs-all, regression, and metric labeling, where the latter is shown to outperform the others. This study differs from our method in that the objects and labels that are being matched in here are entire reviews. In contrast, we utilize metric labeling for comparing two sentence where the object and label nodes are the words of sentences.

3   Embedding Sentences into Graphs

Tackling the QA problem as an instance of graph matching requires the embedding of questions into directed weighted graphs. In this section, we explain the general case of embedding sentences into graphs that is also applicable to the embedding of questions. Each word in a sentence has a corresponding node in the graph whose features are the POS and NE tags of the word, the vector representation of the word in a language model, and the word itself. We used the POS tagger and the NE recognizer of the Stanford CoreNLP toolkit [15] for obtaining POS and NE tags. In order to obtain the vector representation of words, we used the English language model of Mikolov et al. [17] which is trained by *word2vec* system using the Google news dataset. The language model contains three million words, each of which are represented by 300-dimensional vectors.

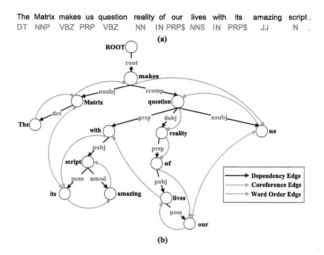

(b)

Fig. 1. Graph representation of a sentence. (a) Acronyms written in red are the POS tags of the corresponding words such as *JJ* and *PRP$* representing adjective and possessive pronoun, respectively. (b) Acronyms written on dependency edges represent the type of relation between two endpoints of the edge such as *amod* and *ccomp* standing for adjectival modifier and clausal complement, respectively. Different edge types are color coded.

Relations among words are represented by directed edges in the graph which is defined as one of the following three types: *word order edges, dependency edges,* and *coreference edges.* Words that follow each other in the sentence are connected by word order edges that point from a word to the next. Edges that are obtained by the dependency parse tree of the sentence are used as dependency edges. Coreferencing words are connected with bi-directional coreference edges. Note that, there can be more than one edge between a pair of nodes. We used the Stanford dependency parser and the Stanford coreference resolution system of the Stanford CoreNLP package [15] for obtaining the aforementioned relations. Graph representation of a sentence is shown in Fig. 1. Distinct edge weights can be associated with types of edges since each type represents a different relation which are needed to be learned empirically.

4   Graph Based Text Matching

Once the questions are embedded into graphs, questions that are similar to a query can be retrieved from a question/answer dataset by matching corresponding graphs. In order to achieve matching, we utilize the metric labeling approach [12] with its primal-dual realization [7]. The goal is to find a similarity score while obtaining a mapping between the nodes of the two graphs (see Fig. 2). In the rest of the text, we present a primal-dual approximation algorithm that is motivated by the method that we previously introduced in [22] for matching lineage trees of the immune repertoire.

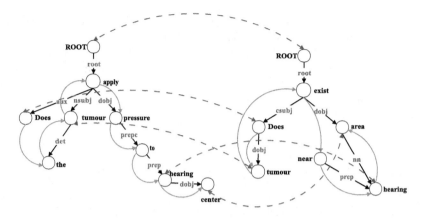

Fig. 2. Matching graph representation of two questions.

We first state the metric labeling problem in its quadratic form as follows.

$$\min \sum_{p\in P}\sum_{a\in L} c_{p,a}\cdot x_{p,a} + \sum_{p\in P}\sum_{q\in P} w_{p,q}\sum_{a\in L}\sum_{b\in L} d_{a,b}\cdot x_{p,a}\cdot x_{q,b}$$

$$\text{s.t.}\ \sum_{a\in L} x_{p,a} = 1, \qquad\qquad \forall p\in P \qquad\qquad (1)$$

$$x_{p,a} \in \{0,1\}, \qquad\qquad p\in P, a\in L$$

where $c_{p,a}$ is the cost of assigning a word p in the query to another word a in a dataset sentence, $w_{p,q}$ is the strength of relation between words p and q of the query sentence, and $d_{a,b}$ is the distance between words a and b of the dataset sentence. Adapting the formulation (1) to graph based text matching requires us to define proper similarity and distance measures for the $c, w,$ and d terms.

4.1 Assignment Cost

In graph representations of sentences, the cost $c_{p,a}$ of assigning word p to word a is defined as a function of four parameters: the vector representation of the words, the dictionary features of the words, their POS tags, and the NE tags. First, the cosine distance between the vector representation of two words that are obtained from the language model is used as a similarity measure between the two words. Next, another similarity score that is obtained according to the dictionary features of the words is used, such as assigning higher similarity to two words that are synonyms or hyponyms. To calculate the similarity score among word pairs, we utilized *WordNet* [18] which is a lexical database for English that groups words into sets of cognitive synonyms by using a graph structure. We used the method of Do et al. [6] among several similarity metrics defined over WordNet. Finally, we take the POS and NE tags into account while determining the similarity score. This is especially important in distinguishing two words that are same but used within different contexts. The following two sentences are an example of such a case over the word *rolling*: "Rolling her eyes, she started to walk away" vs "The Rolling Stones was his favorite rock band". The (POS, NE) tag tuple for the word "Rolling" will be (verb, none) in the first sentence while it is (proper noun, organization) in the second. Even though the vector representation will be the same for both words and WordNet will consider the two to be identical, their similarity score will be set low by incorporating POS and NE tags.

4.2   Separation Cost

The second sum in the objective function of (1), i.e., separation cost, requires defining a relation among nodes over the graph representation of sentences. We need to consider several aspects while defining this relation which in turn will be used for defining a distance measure. First, note that there are three types of edges in the text graphs. Second, it is possible to have an edge of each type (making a total of up to three edges) between any two nodes in the graph. Third, the relation between any two words is proportional to the number of direct edges between them. Finally, for pairs of words that are not directly connected by any edge, the relation should be defined in terms of the path connecting the two. We denote the strength of relation among words p and q as $w_{p,q}$. In order to calculate the weight matrix w, we start with initializing it to zero and then run the following path tracking procedure (denoted *trackPath()*) for each node in the graph. Starting from the node representing word p, we follow the outgoing edges recursively while labeling edges as visited to avoid cycles. We keep a counter for each path denoting the number of hops made since the starting node. For each node q that is encountered on the path, we increase the weight $w_{p,q}$ by the weight coefficient of the type of incoming edge to node q divided by the number of hops made up to this point. Note that, the contribution of edges that are farther in the path will be relatively low due to the scaling of weights by the hop counter. Once we calculate the weight matrix, we set the distance term $d_{p,q}$ as the reciprocal of $w_{p,q}$ for all nodes p, q of the graph.

4.3   The Algorithm

A primal-dual realization of the metric labeling problem that is adapted to text matching is presented in Algorithm 1. The procedure takes as input the graph representations P and L of query and dataset sentences. The initialization phase of the algorithm (lines 1–10) starts with setting the indicator variables x_{pa} to zero which denotes the probability of mapping the query sentence word p to dataset sentence word q (line 1). Next, the costs of assigning query graph nodes to dataset graph nodes $c_{p,a}$ are calculated as a function of the word feature coefficients $\rho_{w2v}, \rho_{NE}, \rho_{POS}$, and ρ_{wn} (line 2). Then, the strength of relation between node pairs w for both graphs P and L are calculated using the *trackPath()* procedure as a function of the number of hops and the edge weight coefficients $\sigma_{coref}, \sigma_{wordOrder}, \sigma_{dependency}$ (lines 4–7). This is followed by the distance measure $d_{a,b}$ which is only calculated for the dataset graph L (line 8). The procedure

---

Algorithm 1. A primal-dual approximation algorithm for the metric labeling problem to match sentences

---

  procedure ML-Primal-Dual-for-Text(P, L)
1: $\forall\, p, q \in P, a \in L : x_{pa} \leftarrow 0$
2: $\forall\, p \in P, a \in L : c_{pa} \leftarrow similarity(p, a, \rho_{w2v}, \rho_{ne}, \rho_{pos}, \rho_{wordNet})$
3: $\forall\, p, q \in P, a, b \in L : w_{pq} \leftarrow 0, w_{ab} \leftarrow 0$
4: **for** $\forall\, p \in P, a \in L$ **do**
5:   $trackPath(p, hops, \sigma_{coref}, \sigma_{wordOrder}, \sigma_{dependency})$
6:   $trackPath(a, hops, \sigma_{coref}, \sigma_{wordOrder}, \sigma_{dependency})$
7: **end for**
8: $\forall\, a, b \in L : d_{a,b} \leftarrow 1/w_{a,b}$
9: $\forall\, p \in P, a \in L : \phi(p, a) \leftarrow c_{pa}$
10: $\mathcal{O} \leftarrow P$
11: **while** $\mathcal{O} \neq \emptyset$ **do**
12:   Find $p \in \mathcal{O}$ that minimizes $\phi(p, a)$ for some $a \in L$
13:   $x_{pa} \leftarrow 1$
14:   $\mathcal{O} \leftarrow \mathcal{O} \setminus \{p\}$
15:   $\forall q \in \mathcal{O}, b \in L \setminus \{a\} : \phi(q, b) = \phi(q, b) + w_{pq} \cdot d_{ab}$
16: **end while**
17: **return** $\sum\limits_{p \in P, a \in L} \phi(p, a)$

---

further defines an adjusted assignment cost function ϕ which decides the mapping of nodes between P and L. The value of $\phi(p, a)$ is initially set to be the assignment cost of p to a (line 9). At each iteration of the loop in lines 11–16, the algorithm makes an assignment for the word pair (p, a) that minimizes the adjusted assignment cost function ϕ (lines 12–13). Before proceeding to the next iteration, ϕ function is updated for each of the query sentence nodes that are still unassigned by an amount of separation cost with respect to the recently assigned node (line 15). The algorithm terminates once all of the nodes in the query sentence are assigned. The outcome of the algorithm is the similarity score between the two graphs which is simply the summation of the adjusted assignment costs.

Proposition 1. *Given $|P| = n$, $|L| = m$, and trackPath() is called with r as the number of hops, running time complexity of Algorithm 1 is $O(n^2 m + rn^2 + rm^2)$.*

Proof. The initializations made in lines 1–3 and 8–10 take $O(n^2 + m^2 + nm)$ time. The time complexity analysis for the lines 4–7 is as follows. The running time of the *trackPath()* procedure is a function of r, i.e., the number of maximum hops that is to be followed for each path, and the number of paths that can be pursued beginning from the starting node. First, note that in a graph with k nodes, there exist $k - 1$ dependency edges since the output of the dependency parser is a tree, and $k - 1$ word order edges between the k nodes. We ignore the contribution of coreference edges in our calculations since it can practically be taken as constant due to the fact that sentences do not contain more than a few coreferencing words. The worst case running time occurs when the parse tree of the sentence is star shaped. Considering the result of graph embedding, which connects the leaf nodes with word order edges, the maximum running time of the *trackPath()* procedure will be observed when the procedure is called for the root of the parse tree. Specifically, the root node will recursively call the *trackPath()* procedure with $r-1$ as the hop count for each of its neighboring nodes, and each function will terminate after at most $r-1$ hops over edges. Thus, the worst case running time of the path tracking procedure is $O(r \cdot k)$ for a graph containing k nodes. Running the path tracking procedure for each node on the query and the dataset graphs, the initialization of the weight matrix takes $O(rn^2 + rm^2)$ time in total. The while loop is executed n times, and lines 11, 13, and 14 each take $O(n)$ time. Making an aggregate analysis for lines 12 and 15, we get $O(n^2)$ and $O(n^2m)$ time, respectively. Thus, the asymptotic running time of Algorithm 1 is $O(n^2m + rn^2 + rm^2)$.

4.4   Learning Coefficients

Several parameters are required to be learned by the algorithm. These parameters include the coefficients $\rho_{w2v}, \rho_{ne}, \rho_{pos}, \rho_{wordNet}$ that account for the contribution of word features word2vec, WordNet similarity, NE, and POS tags to the assignment cost, and $\sigma_{coref}, \sigma_{wordOrder}, \sigma_{dependency}$ that define the weights of the edge types for the calculation of distance measures. Learning these parameters is a non-trivial task due to the time complexity of the graph based matching methods. Nevertheless, the efficiency of the primal-dual algorithm presented in Algorithm 1 makes it possible to learn the parameters in a reasonable time frame. In order to train the system, we used grid search.

5   Experiments

We evaluate our graph based text matching and question answering method on a combination of the Visual Question Answering (VQA) dataset [1] and the Microsoft Common Objects in Context (MS COCO) [13] dataset. The VQA is a visual question answering dataset containing three multiple question and answer pairs for over $80\,K$ images in its training set where the images are sampled from the MS COCO dataset. The MS COCO is an image dataset that is generally used for evaluating image recognition, segmentation, and captioning methods. It provides five captions for each of the images in the dataset along with other information. In our experiments, we obtained question-answer pairs from the VQA dataset and image captions from the MS COCO dataset. We discarded any further information that is provided by the datasets including the images and image features. A set of sample question-answer pairs and captions are shown in Table 1 along with their corresponding image.

Our experimental setup consists of three steps: preprocessing of the dataset, training the parameters, and question answering. At the preprocessing stage, we embedded the questions and captions into graphs as described in Sect. 3. In order to train the parameters, we subsampled questions, answers, and captions corresponding to $5\,K$ images and run the question answering experiment on this

Table 1. Sample captions and question-answer pairs for a corresponding image from our experimental setup. Note that, images are not used as input in the experiments.

Image Captions:
- a stop sign directs pedestrians as a train travels by.
- a stop sign sits in the middle of the forest.
- a stop sign sitting near some tall trees.
- a man leans out of a vehicle near a short stop sign in a forest.
- a stop sign is shown close to the ground.

Question/Answer pairs:
- are there people? - yes
- what kind of sign is in the picture? - stop
- is this a stop sign for a train? - yes

small dataset with various parameter values. We investigated the parameter space using grid search in a coarse to fine level of detail. Specifically, we first run the experiment at a coarse level with equally distributed intervals of parameters. According to the evaluation results of the first level, we continued running the experiment to further levels by finely dividing the region of the parameter space with the highest success rate until reaching a local maximum.

The question answering task is performed as follows. First, the system is loaded with the graph representations of the questions, answers, and the captions. Questions are sorted into groups according to the question category that is provided by the VQA dataset, such as "what" or "why". Question types for which there exist less than 100 questions in the dataset are grouped together into a category called "others". Next, the question answering experiment is evaluated using the leave-one-out approach with k-nearest neighbor (k-NN) algorithm. That is, questions and captions corresponding to an image are withdrawn from the dataset which is referred as the query. The goal is to find a correct answer to each of the three questions that are asked for the query image. We first rule out the question types that do not match with the question type of the query which we take as its first word. Next, we run the graph based text matching between the query and the questions in the dataset to obtain the most similar n questions. Using graph matching, we further match the image captions of the query with the captions of images corresponding to the most similar n questions. Among the resulting similarity scores, we choose the k nearest neighbors out of the n questions. Finally, we decide on the answer using a voting scheme over the multiple choices of the query through the answers to the k most similar

Table 2. Accuracy of the methods (in percentage) for various k-NN values. Results are reported for the cases when only the questions are used as well as questions and captions are used in combination.

Method	Questions only			Questions & Captions		
	1-kNN	2-kNN	3-kNN	1-kNN	2-kNN	3-kNN
3-grams	35.73	52.79	67.00	37.69	54.07	67.00
Multi-layer perceptron [1]	53.68			59.85		
Our method	36.43	54.65	69.38	37.85	55.80	69.61

question-answer pairs. We have experimentally chosen n and k values to be 100 and 50, respectively.

We took the n-gram based matching as the baseline to compare to our graph based text matching method. We used n-grams up to 3-grams for describing the questions and captions. The l_1 distance is then used as the similarity measure between n-grams, which is normalized according to the number of tokens in each vector. Results are reported in Table 2.

We evaluated the accuracy of the methods for two cases. First, the answers are decided based on the answers of the similar questions. Second, we further take the image captions into account and refine the set of similar questions by matching the image captions. As shown in Table 2, including the caption information to the question answering process slightly increases the accuracy of the system. This indicates that the questions are informative enough to provide a correct answer for the multiple choice questions. Also note that, the correct answer is within the 3 highly rated answers (3-NN) 70% of the time among the 18 multiple choices provided for each question for both the n-gram method and our approach. Our graph matching algorithm achieved better performance rates compared to the n-gram with the current set of parameters that we used. However, both methods perform relatively poorly when contrasted to the multi-layer perceptron based method presented in [1]. We note that, our method's performance has the potential to be improved by training the algorithm with a larger set of question-answer pairs.

6   Conclusions and Future Work

In this paper, we presented a graph based text matching approach applied to the question answering problem. We first provided an embedding method to represent sentences in natural language by directed weighted graphs. Next, we formulated the text matching problem as an instance of the metric labeling problem. Finally, we provided the application of the primal-dual approximation algorithm for metric labeling to the text matching. The experimental evaluation of our method on the VQA dataset shows the success of graph based matching over the n-gram based matching.

One limitation of the proposed system emanates from using the k-nearest neighbor method to decide about the answer. Although the primal dual approximation of metric labeling is efficient, the approach is not scalable considering the growing sizes of knowledge bases. In the future, we would like to investigate the utility of other machine learning algorithms. For instance, a kernel learning approach such as SVM [5] can be used in a multi-class setting with question types defining classes, in order to identify a subset of dataset questions with salient properties (that is, support vectors). Then, we can use only these identified questions to compare with the query to reduce the running time. Our future goal is to investigate the feasibility of applying the primal dual approximation of the metric labeling as the graph kernel for the SVM. If this hypothesis fails, we would like to investigate the use of clustering algorithms such as K-medoids [24] on dataset questions to identify data-driven clusters of questions and their representative members.

References

1. Antol, S., Agrawal, A., Lu, J., Mitchell, M., Batra, D., Lawrence Zitnick, C., Parikh, D.: VQA: visual question answering. In: Proceedings of the IEEE International Conference on Computer Vision, pp. 2425–2433 (2015)
2. Brill, E., Dumais, S., Banko, M.: An analysis of the AskMSR question-answering system. In: Proceedings of the ACL-02 Conference on Empirical methods in Natural Language Processing, vol. 10, pp. 257–264. Association for Computational Linguistics (2002)
3. Cavnar, W.B., Trenkle, J.M.: N-gram based text categorization. In: 3rd Annual Symposium on Document Analysis and Information Retrieval Proceedings of SDAIR 1994, pp. 161–175 (1994)
4. Clark, J.H., González-Brenes, J.P.: Coreference resolution: current trends and future directions. In: Language and Statistics II Literature Review, p. 14 (2008)
5. Cortes, C., Vapnik, V.: Support-vector networks. Mach. Learn. **20**(3), 273–297 (1995)
6. Do, Q., Roth, D., Sammons, M., Tu, Y., Vydiswaran, V.: Robust, light-weight approaches to compute lexical similarity. Computer Science Research and Technical Reports, University of Illinois (2009)
7. Goemans, M.X., Williamson, D.P.: The primal-dual method for approximation algorithms and its application to network design problems. In: Hochbaum, D.S. (ed.) Approximation Algorithms for NP-hard Problems, pp. 144–191. PWS Publishing Co., Boston (1997)
8. Green Jr., B.F., Wolf, A.K., Chomsky, C., Laughery, K.: Baseball: an automatic question-answerer. In: Papers presented at the May 9–11, 1961, western joint IRE-AIEE-ACM computer conference, pp. 219–224. ACM (1961)
9. Haghighi, A.D., Ng, A.Y., Manning, C.D.: Robust textual inference via graph matching. In: Proceedings of the conference on Human Language Technology and Empirical Methods in Natural Language Processing, pp. 387–394. Association for Computational Linguistics (2005)
10. Horn, L.R., Kato, Y., Negation, P.: Syntactic and Semantic Perspectives. Oxford University Press, Oxford (2000)

11. Iyyer, M., Boyd-Graber, J.L., Claudino, L.M.B., Socher, R., Daumé III., H.: A neural network for factoid question answering over paragraphs. In: EMNLP, pp. 633–644 (2014)
12. Kleinberg, J., Tardos, É.: Approximation algorithms for classification problems with pairwise relationships: Metric labeling and markov random fields. J. ACM 49(5), 616–639 (2002)
13. Lin, T.-Y., Maire, M., Belongie, S., Hays, J., Perona, P., Ramanan, D., Dollár, P., Zitnick, C.L.: Microsoft COCO: common objects in context. In: Fleet, D., Pajdla, T., Schiele, B., Tuytelaars, T. (eds.) ECCV 2014. LNCS, vol. 8693, pp. 740–755. Springer, Cham (2014). doi:10.1007/978-3-319-10602-1_48
14. Manning, C.D., Raghavan, P., Schütze, H.: Introduction to information retrieval, vol. 1. Cambridge University Press, Cambridge (2008)
15. Manning, C.D., Surdeanu, M., Bauer, J., Finkel, J.R., Bethard, S., McClosky, D.: The stanford corenlp natural language processing toolkit. In: ACL (System Demonstrations), pp. 55–60 (2014)
16. Marquez, L.: Machine learning and natural language processing. Technical Report LSI-00-45-R, Departament de Llenguatges i Sistemes Informatics (LSI), Universitat Politecnica de Catalunya (UPC), Barcelona, Spain (2000)
17. Mikolov, T., Chen, K., Corrado, G., Dean, J.: Efficient estimation of word representations in vector space. arXiv preprint arXiv:1301.3781 (2013)
18. Miller, G.A.: Wordnet: a lexical database for english. Commun. ACM 38(11), 39–41 (1995)
19. Mitkov, R., Ha, L.A.: Computer-aided generation of multiple-choice tests. In: Proceedings of the HLT-NAACL 2003 workshop on Building Educational Applications Using Natural Language Processing, vol. 2, pp. 17–22. Association for Computational Linguistics (2003)
20. Moschitti, A.: Efficient convolution kernels for dependency and constituent syntactic trees. In: Fürnkranz, J., Scheffer, T., Spiliopoulou, M. (eds.) ECML 2006. LNCS, vol. 4212, pp. 318–329. Springer, Heidelberg (2006). doi:10.1007/11871842_32
21. Okita, T.: Local graph matching with active learning for recognizing inference in text at ntcir-10. In: NTCIR 10 Conference, pp. 499–506 (2013)
22. Osmanlıoğlu, Y., Ontañón, S., Hershberg, U., Shokoufandeh, A.: Efficient approximation of labeling problems with applications to immune repertoire analysis. In: 23rd International Conference on Patter Recognition, ICPR 2016 (2016)
23. Pang, B., Lee, L.: Seeing stars: exploiting class relationships for sentiment categorization with respect to rating scales. In: Proceedings of the 43rd Annual Meeting on Association for Computational Linguistics, ACL 2005, pp. 115–124, Stroudsburg, PA, USA. Association for Computational Linguistics (2005)
24. Park, H.-S., Jun, C.-H.: A simple and fast algorithm for k-medoids clustering. Expert Syst. Appl. 36(2), 3336–3341 (2009)
25. Ramachandran, L., Gehringer, E.F.: Graph-structures matching for review relevance identification. In: Graph-Based Methods for Natural Language Processing, p. 53 (2013)
26. Roth, D., Kao, G.K., Li, X., Nagarajan, R., Punyakanok, V., Rizzolo, N., Yih, W., Alm, C.O., Moran, L.G.: Learning components for a question-answering system. In: TREC (2001)
27. Sizov, G., Öztürk, P.: Automatic extraction of reasoning chains from textual reports. In: Graph-Based Methods for Natural Language Processing, p. 61 (2013)
28. Turing, A.M.: Computing machinery and intelligence. Mind 59(236), 433–460 (1950)

Hierarchical Bayesian Multiple Kernel Learning Based Feature Fusion for Action Recognition

Wen Sun[1], Chunfeng Yuan[1(✉)], Pei Wang[1], Shuang Yang[1], Weiming Hu[1], and Zhaoquan Cai[2]

[1] CAS Center for Excellence in Brain Science and Intelligence Technology, National Laboratory of Pattern Recognition, Institute of Automation, Chinese Academy of Sciences, Beijing, China
{wen.sun,cfyuan,pei.wang,syang,wmhu}@nlpr.ia.ac.cn
[2] Huizhou University, Huizhou, Guangdong, China
gd0752888@126.com

Abstract. Human action recognition is an area with increasing significance and has attracted much research attention over these years. Fusing multiple features is intuitively an appropriate way to better recognize actions in videos, as single type of features is not able to capture the visual characteristics sufficiently. However, most of the existing fusion methods used for action recognition fail to measure the contributions of different features and may not guarantee the performance improvement over the individual features. In this paper, we propose a new *Hierarchical Bayesian Multiple Kernel Learning* (HB-MKL) model to effectively fuse diverse types of features for action recognition. The model is able to adaptively evaluate the optimal weights of the base kernels constructed from different features to form a composite kernel. We evaluate the effectiveness of our method with the complementary features capturing both appearance and motion information from the videos on challenging human action datasets, and the experimental results demonstrate the potential of HB-MKL for action recognition.

Keywords: Action recognition · Feature fusion · Multiple kernel learning

1 Introduction

Action recognition is an active research area in computer vision motivated by the promise of applications in broad domains such as intelligent surveillance, human-computer interaction and video retrieval. However, the task is still challenging due to the variations in action performances, background clutter, illumination changes, camera movements and occlusions.

The previous researches [1–7] in the literature have paid more attention to designing descriptive features which are specific to action recognition and a large number of features are available now for this task. It is an intuitive way to integrate diverse types of informative features instead of a single one to improve

© Springer International Publishing AG 2017
F. Schwenker and S. Scherer (Eds.): MPRSS 2016, LNAI 10183, pp. 85–97, 2017.
DOI: 10.1007/978-3-319-59259-6_8

the recognition performance. However, the existing action recognition algorithms [8,9] usually employ the simple combination of different features. The most common method is the feature-level fusion which concatenates all the feature vectors together into one single feature vector. A drawback of the method is the high dimensionality of the final concatenated vector, since the efficiency of the method drops exponentially as the dimensionality increases. Another feasible solution is the kernel-level fusion. For instance, the multi-channel approach proposed in [10] simply takes the multiplication of the kernels. Nevertheless, the method cannot guarantee the performance improvement over the individual features. It is worth noting that both methods do not consider the relative importance of the candidate features and this leads to a meaningless combination. Therefore, it requires to formulate a combination method that is able to evaluate the relative contributions of different feature representations and utilize such information to gain enhanced classification performance.

In this paper, we propose a new *Hierarchical Bayesian Multiple Kernel Learning* (HB-MKL) framework to deal with feature fusion problem for action recognition. We first formulate the multiple kernel learning problem as a decision function based on a weighted linear combination of the base kernels, and then develop a hierarchical Bayesian framework with three layers to solve this problem. Specifically, the bottom layer consists of the parameters in the decision function. On the middle layer, the priors of Gaussian distribution family are placed on the parameters of the decision function. Especially, the prior on the kernel weight is set by a half-normal distribution, which has the advantage of interpretability due to the only nonnegative restriction in nature. The top layer is composed of the hyper-priors, invoked on the parameters of the priors at the level below. Gamma distribution is employed to take the advantage of the conjugacy and non-informativeness. The non-informativeness ensures that the learnt model parameters are intrinsic to the data. The model is established in a fully conjugate manner, offering the probability of efficient inference. Therefore, we derive a variational approximation for inference. After evaluating the optimal weights of the base kernels using the framework above, we derive the composite kernel. Finally, an SVM classifier is trained using the learnt optimal combined kernel. We apply the above model to the feature fusion problem in the field of action recognition, where no such attempts have been made before to the best of our knowledge. We conduct a set of experiments for better illustration and comparison on several public action datasets. The experimental results demonstrate the effectiveness of our method and provide some insight on the contributions of different features for action recognition.

The main contributions of this work can be summarized as follows. First, a new framework of hierarchical Bayesian multiple kernel learning is designed. The half-normal distribution prior placed on the base kernel weights makes them nonnegative without any other constraints, which exactly meets the actual requirements and has good interpretability. Second, instead of conventional simple fusion of multiple features used in action recognition, we propose to apply the HB-MKL based feature fusion method to action recognition, which can learn the optimal combination of multiple features automatically. Third, we carry out a set of experiments on three datasets, and the experimental results demonstrate the efficiency of the proposed method. It is worth mentioning that the valuable results of the feature weights learnt by our method give some insight on how each feature contributes to recognizing an action.

2   Related Work

In this section, we give a brief overview of the related work on three aspects: discriminative features for action recognition, feature fusion methods and multiple kernel learning algorithms.

Various classical video feature descriptors are proposed in previous work including HOG, HOF [1], MBH [2] and some spatio-temporal extensions of image descriptors, such as 3D-SIFT [3], HOG3D [4] and extended SURF [5]. Moreover, trajectory features are also popular descriptors. In [6], human actions are represented using sparse SIFT-based trajectory. Wang et al. [7] introduce an approach to combine dense sampling with feature tracking, and extract robust features along the trajectories.

Realizing it is not enough to describe videos using homogeneous descriptor, some researchers try to fuse heterogeneous descriptors to construct more discriminative classifiers. However, most of the existing algorithms combine multiple features in an easy way. Tian et al. [8] combine the histogram of MHI and Haar wavelet transform of MHI at the feature-level. They use the straightforward concatenation of the features as the combined feature representation, which is a higher dimensional vector. Ullah et al. [9] use a multi-channel approach proposed in [10] to integrate feature representations, which takes the multiplication of the feature kernels in nature. The method can be regarded as a combination at the kernel-level using fixed rules without additional parameters. However, the above mentioned methods do not take into account the contribution of different features and hence cannot make better use of the multiple features. In this paper, we employ Multiple Kernel Learning (MKL) to informatively combine diverse features for action recognition.

Many variants of MKL have been proposed in the previous work. In this paper, we consider MKL with a weighted linear combination of the base kernels under a Bayesian framework. The existing Bayesian MKL methods differ in the prior assumptions on the kernel weights. Girolami et al. [11] present a Bayesian model for regression and classification problems by employing a Dirichlet prior on the kernel weighting coefficients. Damoulas et al. [12] use a similar model with the same prior distribution assumptions and extend the model for multiclass problem. Moreover, they apply the approach to protein fold recognition and remote homology detection problems to prove the validity of the method. Gönen [13] presents an efficient MKL algorithm by assuming the kernel weights to be normally distributed. In this paper, we introduce a half-normal distribution on the kernel weights. Compared with the normal distribution prior, the half-normal distribution ensures that the kernel weights are nonnegative and hence it produces a more meaningful combination of kernels.

3   Hierarchical Bayesian Multiple Kernel Learning for Action Recognition

In this section, we first introduce the heterogeneous and complementary features used to sufficiently represent the actions in videos. Then we introduce the detailed HB-MKL algorithm and its inference. Finally, we apply HB-MKL to effectively fuse the obtained multiple features for action recognition.

3.1   Multiple Features for Action Representation

In this paper, we use the state-of-the-art improved dense trajectory features [14] for action representation. We first extract the trajectories by densely sampling feature points in each frame and tracking them in the video based on displacement information from the optical flow field. Subsequently, we compute the trajectory-aligned descriptors (i.e., Trajectory, HOF, HOG and MBH) within a space-time volume along the trajectories.

It is worth noting that the extracted features are complementary in describing action sequences by capturing both static appearance and dynamic motion information. The trajectory descriptor is a concatenation of normalized displacement vectors which describe the motion of the trajectories. HOF captures the motion information based on the orientation of optical flow, whereas HOG calculates the histograms of oriented gradients which measure the static appearance information. MBH (motion boundary histogram) encodes relative motion information by computing derivatives separately for the horizontal and vertical components of the optical flow.

Once we obtain the features above, we encode them using both Bag of Features (BOF) and Fisher Vector (FV) [15] approaches to achieve the final video sequence representations. Using one of these two strategies, each video is represented by four kinds of features which characterize complementary information of the video sequence.

3.2   Hierarchical Bayesian Multiple Kernel Learning

In order to formulate a better combination of the obtained multiple features, we propose a HB-MKL model for feature fusion. First, we formulate the MKL for multi-class classification problem as described below.

Consider N independent and identically distributed training instances $\{\mathbf{x}_i\}_{i=1}^N$, where each data instance has P feature representations $\mathbf{x}_i = \{\mathbf{x}_i^m\}_{m=1}^P$. In this paper, we consider a combined kernel which fuses different kinds of feature kernels in a linear way as follows:

$$K_e(\mathbf{x}_i, \mathbf{x}_j) = \sum_{m=1}^P e_m K_m(\mathbf{x}_i^m, \mathbf{x}_j^m), \tag{1}$$

where K_m is the base kernel calculating a similarity metric between videos with respect to the m-th feature, e_m is the corresponding kernel weight indicating the m-th base kernel's contribution and significance, and K_e is the composite kernel that finally measures the overall similarity between two videos. Based on the obtained composite kernel K_e, the decision function for a test instance $\mathbf{x}_*$ with respect to action class c can be written as:

$$f^c(\mathbf{x}_*) = \sum_{i=1}^{N} a_c^i K_e(\mathbf{x}_i, \mathbf{x}_*) + b_c, \quad c = 1, \cdots, K, \tag{2}$$

where K is the number of the action classes, a_c^i denotes the weight assigned to the i-th training instance for the c-th action class, and b_c is the bias for the c-th action class.

We then propose a hierarchical probabilistic model to solve the above multi-class multiple kernel learning problem in a Bayesian manner. Specifically, we impose that the kernel weight e_m is sampled from a half-normal distribution with precision ω_m, which ensures that the kernel weights are non-negative without any other constraints. The training instance weight a_c^i and the bias b_c are placed by two zero-mean Gaussian distributions with precisions λ_c^i and γ_c, respectively. Thus according to the decision function, the classification score f_i^c is generated from a Gaussian distribution with the mean $\mathbf{e}^{\mathrm{T}} \mathbf{a}_c^{\mathrm{T}} \mathbf{k}_{m,i} + b_c$ and precision 1. Given the classification score f_i^c, the corresponding class label y_i^c is simply obtained by setting a threshold ν.

Table 1. List of notations

Notations	Dimensions	Representations
$\{\mathbf{K}_m\}_{m=1}^{P}$	$N \times N$	Base kernel matrices
$\mathbf{A}$	$N \times K$	Training instance weight matrix
$\boldsymbol{\lambda}$	$N \times K$	Priors for training instance weight matrix
$\mathbf{e}$	P	Kernel weight vector
$\boldsymbol{\omega}$	P	Priors for kernel weight vector
$\mathbf{b}$	K	Bias vector
$\boldsymbol{\gamma}$	K	Priors for bias vector
$\mathbf{F}$	$K \times N$	Classification score matrix
$\mathbf{Y}$	$K \times N$	Class label matrix

Finally, three non-informative Gamma distributions with different shape and scale parameters are placed on the precisions ω_m, λ_c^i and γ_c of Gaussian distributions respectively. On one hand, the parameters of Gamma distribution are in general non-informative and thus the learnt kernel weights, training instance weights, and biases are intrinsic to the data without prior knowledge assumptions. On the other hand, the above hierarchical probabilistic model is constructed in the conjugate exponential family, and therefore inference can be implemented via variational Bayesian or Gibbs-sampling analysis, with analytic update equations. The variables mentioned above correspond to one instance with respect to one action class. The vector or matrix forms of these variables corresponding to all the training instances are listed in Table 1 for clarity. Actually, the superscripts and subscripts in the notations a_c^i, λ_c^i, f_i^c, y_i^c denote the row and column indexes of their matrices, respectively.

With these parametric definitions, the probabilistic graphical model of our HB-MKL framework for multi-class classification is illustrated in Fig. 1. Corresponding to the three layers in the graphical model, the proposed HB-MKL is expressed in the following three groups of formulations in summary. On the bottom layer, the classification score of the instance with respect to action class c is expressed as:

$$f_i^c|b_c, e, a_c, k_{m,i} \sim \mathcal{N}(f_i^c; e^{\mathrm{T}} a_c^{\mathrm{T}} k_{m,i} + b_c, 1)$$
$$y_i^c|f_i^c \sim \delta(f_i^c y_i^c > \nu), \tag{3}$$

where $\mathcal{N}(\cdot; \mu, \Sigma)$ denotes a Gaussian distribution with the mean vector μ and the covariance matrix Σ, and $\delta(\cdot)$ represents the Kronecker delta function.

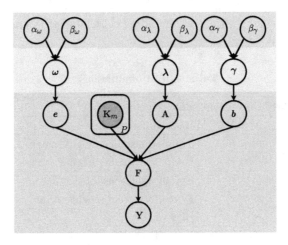

Fig. 1. Graphical model of hierarchical Bayesian multiple kernel learning

On the middle layer, the half-normal distribution and Gaussian distribution are placed on the parameters of the decision function, which are expressed as:

$$e_m|\omega_m \sim \mathcal{N}^+(e_m; 0, \omega_m^{-1})$$
$$a_c^i|\lambda_c^i \sim \mathcal{N}(a_c^i; 0, (\lambda_c^i)^{-1})$$
$$b_c|\gamma_c \sim \mathcal{N}(b_c; 0, \gamma_c^{-1}), \tag{4}$$

where $\mathcal{N}^+(\cdot; 0, \Sigma)$ denotes a half-normal distribution with the mean vector 0 and the covariance matrix Σ.

On the top layer, non-informative gamma hyper-priors are placed on ω_m, λ_c^i and γ_c as follows:

$$\omega_m \sim \mathcal{G}(\omega_m; \alpha_\omega, \beta_\omega)$$
$$\lambda_c^i \sim \mathcal{G}(\lambda_c^i; \alpha_\lambda, \beta_\lambda)$$
$$\gamma_c \sim \mathcal{G}(\gamma_c; \alpha_\gamma, \beta_\gamma), \tag{5}$$

where $\mathcal{G}(\cdot; \alpha, \beta)$ denotes a Gamma distribution with the shape and scale parameters α and β.

3.3 Variational Inference

In order to perform efficient processing, we derive variational approximation methodology for inference. The variational method [16], offers a lower bound on the model evidence using an ensemble of factored posteriors to approximate the joint parameter posterior distribution. By defining the sets of priors as $\boldsymbol{\Xi} = \{\boldsymbol{\gamma}, \boldsymbol{\lambda}, \boldsymbol{\omega}\}$, hyper-priors as $\boldsymbol{\zeta} = \{\alpha_\gamma, \beta_\gamma, \alpha_\lambda, \beta_\lambda, \alpha_\omega, \beta_\omega\}$, and the remaining variables as $\boldsymbol{\Theta} = \{\mathbf{A}, \boldsymbol{b}, \boldsymbol{e}, \mathbf{F}\}$, the factorable ensemble approximation of the required posterior can be written as

$$p(\boldsymbol{\Theta}, \boldsymbol{\Xi} | \boldsymbol{\zeta}, \{\mathbf{K}_m\}_{m=1}^P, \mathbf{Y}) \approx q(\boldsymbol{\Theta}, \boldsymbol{\Xi}) = q(\boldsymbol{\lambda})q(\mathbf{A})q(\boldsymbol{\omega})q(\boldsymbol{e})q(\boldsymbol{\gamma})q(\boldsymbol{b})q(\mathbf{F}), \tag{6}$$

and each factor in the ensemble can be defined as:

$$q(\boldsymbol{\lambda}) = \prod_{i=1}^N \prod_{c=1}^K \mathcal{G}(\lambda_c^i; \alpha(\lambda_c^i), \beta(\lambda_c^i))$$
$$q(\mathbf{A}) = \prod_{c=1}^K \mathcal{N}(\boldsymbol{a}_c; \mu(\boldsymbol{a}_c), \Sigma(\boldsymbol{a}_c))$$
$$q(\boldsymbol{\omega}) = \prod_{m=1}^P \mathcal{G}(\omega_m; \alpha(\omega_m), \beta(\omega_m))$$
$$q(\boldsymbol{e}) = \mathcal{N}^+(\boldsymbol{e}; \mu(\boldsymbol{e}), \Sigma(\boldsymbol{e}))$$
$$q(\boldsymbol{\gamma}) = \prod_{c=1}^K \mathcal{G}(\gamma_c; \alpha(\gamma_c), \beta(\gamma_c))$$
$$q(\boldsymbol{b}) = \mathcal{N}(\boldsymbol{b}; \mu(\boldsymbol{b}), \Sigma(\boldsymbol{b}))$$
$$q(\mathbf{F}) = \prod_{c=1}^K \prod_{i=1}^N \mathcal{TN}(f_i^c; \mu(f_i^c), \Sigma(f_i^c), \rho(f_i^c)).$$

We can bound the model evidence using Jensen's inequality:

$$\log p(\mathbf{Y} | \boldsymbol{\zeta}, \{\mathbf{K}_m\}_{m=1}^P) \geq$$
$$\mathbb{E}_{q(\boldsymbol{\Theta}, \boldsymbol{\Xi})}[\log p(\mathbf{Y}, \boldsymbol{\Theta}, \boldsymbol{\Xi} | \boldsymbol{\zeta}, \{\mathbf{K}_m\}_{m=1}^P)] - \mathbb{E}_{q(\boldsymbol{\Theta}, \boldsymbol{\Xi})}[\log q(\boldsymbol{\Theta}, \boldsymbol{\Xi})], \tag{7}$$

and optimize it with respect to the distribution in the following form

$$q(\tau) \propto \exp(\mathbb{E}_{q(\{\boldsymbol{\Theta}, \boldsymbol{\Xi}\} \backslash \tau)}[\log p(\mathbf{Y}, \boldsymbol{\Theta}, \boldsymbol{\Xi} | \boldsymbol{\zeta}, \{\mathbf{K}_m\}_{m=1}^P)]). \tag{8}$$

3.4   HB-MKL Based Feature Fusion for Action Recognition

In order to utilize the proposed method for action recognition, we first extract and encode the features described above to get the final video descriptors. When adopting BOF representations, we use RBF-χ^2 kernel [1] to separately calculate the base kernels corresponding to different features. As for FV representations, we compute the base kernels using linear kernel function. After that, we apply the proposed HB-MKL to construct a composite kernel by learning the optimum linear combination of the multiple kernels. Finally, we train a standard SVM classifier with the combined kernel. For all the experiments, the multiclass classification is made using the one-vs-all strategy.

4   Experiments

We evaluate our method on three popular human action datasets: KTH, UCF sports, and HMDB51 datasets.

The **KTH** dataset [17] contains six action classes. The actions are performed several times by 25 subjects under 4 different scenarios. The backgrounds are relatively homogeneous and static in most sequences. We follow the experimental settings in [17] where the videos are divided into a training set (16 subjects) and a test set (9 subjects). For evaluation, the average accuracy over all classes is reported.

The **UCF sports** dataset [18] includes 150 sequences of 10 classes of human actions. The videos are extracted from sports broadcasts which are recorded in unconstrained environments with camera motion and different viewpoints. We apply a leave-one-out cross validation scheme and the evaluation is measured using the average accuracy over all classes.

The **HMDB51** dataset [19] contains a total of 6766 video clips collected from various sources, ranging from digitized movies to YouTube. The videos in the dataset vary in video quality, camera motion, viewpoints and occlusions. In our experiments, we adopt the original experimental setup as in [19] with three train/test splits. The average accuracy over the three splits is reported as the performance measurement.

4.1   Baseline Feature Fusion Methods

To evaluate the performance improvement achieved using HB-MKL, we perform experiments with two baseline feature fusion methods for comparison: concatenation and multi-channel methods. The concatenation method directly concatenates all the feature representations together to form a combined representation. The multi-channel method combines different descriptors as follows [10]:

$$K(\mathbf{x}_i, \mathbf{x}_j) = \exp(-\sum_m \frac{1}{A^m} D(\mathbf{x}_i^m, \mathbf{x}_j^m)), \tag{9}$$

where $D(\mathbf{x}_i^m, \mathbf{x}_j^m)$ is the χ^2 distances defined on histogram representations between videos $\mathbf{x}_i$ and $\mathbf{x}_j$ with respect to channel m. A^m is the normalization factor computed as the average value of χ^2 distances between all the training instances for the m-th channel.

4.2   Comparison of Experimental Results

In order to qualify the effectiveness of our approach, we evaluate the classification accuracies achieved by each of the features alone, as well as feature combination via HB-MKL. The results of these approaches using BOF encoding are shown in Table 2. It is clear that feature fusion using HB-MKL outperforms their uses separately on all the datasets. By combining all the features using HB-MKL, we obtain 95.37% on KTH which is around 1% better than the best single feature, whereas on UCF sports it is around 5%. The improvement is even higher on HMDB51, i.e., around 10%. The results demonstrate that the integration of diverse features using HB-MKL enhances the performance compared with single feature based approach.

Table 2. Performance comparisons of five single feature based approaches as well as three fusion approaches using baseline and HB-MKL

Approaches	KTH(%)	UCF(%)	HMDB51(%)
Trajectory	92.13	82.67	33.27
HOF	94.44	85.33	40.37
HOG	87.96	84.00	28.93
MBHx	93.98	82.67	35.80
MBHy	94.44	82.67	42.16
Concatenation	93.98	78.67	39.65
Multi-channel	94.44	77.33	41.33
HB-MKL	**95.37**	**90.00**	**52.07**

In addition, we also compare our method with the baseline combination methods in Table 2. It can be seen that there is a significant performance gain of our combination method over the baselines. Moreover, we notice that the combinations using baselines can not guarantee the improvement with respect to every single features. In contrast, our method consistently outperforms all single features on all the datasets. The advantage of our feature fusion method over baselines can be attributed to the ability of learning the relative importance of each feature.

We also do a performance comparison using different feature encoding strategies. Table 3 lists the results using both BOF and FV for feature encoding. We notice that the improvement of FV over BOF on the KTH dataset is slightly, whereas it reaches 4.6% on HMDB51. Unexpectedly, the performance of FV is inferior to BOF on UCF sports. Based on this evaluation, we choose the best

Table 3. Comparison of feature encoding strategies using BOF and FV

	BOF	FV
KTH(%)	95.37	**95.83**
UCF(%)	**90.00**	88.00
HMDB51(%)	52.07	**56.67**

performed FV encoding for KTH and HMDB51, and BOF encoding for UCF sports in the rest of the experiments.

We also compare our method with the most recent results reported in the literature on the three datasets in Table 4. On KTH, our method yields better performance than [20]. The work of [20] uses direction-dependent feature pairs to represent actions, and achieves a recognition rate of 95.0%. Zhang et al. [21] report 87.5% on UCF sports by using a simplex-based orientation decomposition descriptor to describe 3D visual features. We further improve their results by 2.5%. On HMDB51, Wu et al. [22] report 56.4% with a VLAD-based video encoding for human action recognition. We achieve 56.7% which is slightly better than theirs. It can be seen that the proposed method achieves a comparable performance to the state-of-the-art approaches.

Table 4. Performance comparisons of our method with the state-of-the-art results

KTH		UCF sports		HMDB51	
Sun et al. [23]	93.1%	Sun et al. [23]	86.6%	Yang et al. [24]	53.9%
Zhang et al. [21]	94.8%	Zhang et al. [21]	87.5%	Wu et al. [22]	56.4%
Veeriahet et al. [25]	94.0%	Lan et al. [26]	83.6%	Shao et al. [27]	49.8%
Wang et al. [28]	94.5%	Wang et al. [28]	86.7%	Liu et al. [29]	51.4%
Sheng et al. [20]	95.0%	Sheng et al. [20]	87.3%	Liu et al. [30]	48.4%
Our method	**95.8%**	**Our method**	**90.0%**	**Our method**	**56.7%**

4.3   Analysis of Feature Weights Learnt by HB-MKL

Table 5 shows the feature weights learnt by HB-MKL in the range [0, 1]. From the table, we can see how each feature contributes to the final decision. It is clearly to see that on KTH, among all the feature representations, HOF plays the dominant role, while HOG tends to have the lowest weight. This reveals that motion-based features of a video are the most informative features for action recognition on KTH. This may be because the variation in appearances between frames is very small on KTH.

As for UCF sports and HMDB51, it can be seen that HOG ranks first, followed by motion-based features. This is probably because both of the datasets contain lots of camera motion which reduces the reliability of motion-based features. Moreover, the UCF sports dataset often involves specific environment and equipment, and hence the appearance-based feature is more important for it. Therefore, it demonstrates that the proposed HB-MKL is able to learn the optimal feature weights from data adaptively.

Table 5. The feature representation weights learnt by HB-MKL

	KTH	UCF	HMDB51
Trajectory	0.19	0.23	0.20
HOF	0.23	0.22	0.21
HOG	0.12	0.25	0.24
MBHx	0.23	0.16	0.17
MBHy	0.23	0.14	0.18

5   Conclusion

In this paper, we have presented an efficient feature fusion framework based on hierarchical Bayesian multiple kernel learning for action recognition. The method is able to integrate different features in an informative way by evaluating the relative importance of every feature and finally learns the optimum kernel combination of the multiple feature kernels. We have carried out a set of experiments on three human action datasets to evaluate the effectiveness of our approach, and the results have demonstrated that the proposed approach generally outperforms the state-of-the-art methods in terms of classification accuracy for action recognition.

Acknowledgments. This work is partly supported by the 973 basic research program of China (Grant No. 2014CB349303), the Natural Science Foundation of China (Grant No. 61472421, U1636218, 61472420, 61370185, 61170193, 61472063), the Strategic Priority Research Program of the CAS (Grant No. XDB02070003), the Natural Science Foundation of Guangdong Province (Grant No. S2013010013432, S2013010015940), and the CAS External cooperation key project.

References

1. Laptev, I., Marszalek, M., Schmid, C., Rozenfeld, B.: Learning realistic human actions from movies. In: CVPR, pp. 1–8 (2008)
2. Dalal, N., Triggs, B., Schmid, C.: Human detection using oriented histograms of flow and appearance. In: ECCV, pp. 428–441 (2006)
3. Scovanner, P., Ali, S., Shah, M.: A 3-dimensional sift descriptor and its application to action recognition. In: ACM MM, pp. 357–360 (2007)
4. Klaser, A., Marszalek, M., Schmid, C.: A spatio-temporal descriptor based on 3d-gradients. In: BMVC (2008)

5. Willems, G., Tuytelaars, T., Gool, L.V.: An efficient dense and scale-invariant spatio-temporal interest point detector. In: ECCV, pp. 650–663 (2008)
6. Sun, J., Wu, X., Yan, S., Cheong, L.F., Chua, T.S., Li, J.: Hierarchical spatio-temporal context modeling for action recognition. In: CVPR, pp. 2004–2011 (2009)
7. Wang, H., Kläser, A., Schmid, C., Liu, C.L.: Action recognition by dense trajectories. In: CVPR, pp. 3169–3176 (2011)
8. Tian, Y., Cao, L., Liu, Z., Zhang, Z.: Hierarchical filtered motion for action recognition in crowded videos. IEEE Trans. Syst. Man Cybern. Part C: Appl. Rev. **42**, 313–323 (2012)
9. Ullah, M.M., Parizi, S.N., Laptev, I.: Improving bag-of-features action recognition with non-local cues. In: BMVC, pp. 95.1–95.11 (2010)
10. Zhang, J., Marszałek, M., Lazebnik, S., Schmid, C.: Local features and kernels for classification of texture and object categories: a comprehensive study. IJCV **73**, 213–238 (2007)
11. Girolami, M., Rogers, S.: Hierarchic Bayesian models for kernel learning. In: ICML, pp. 241–248 (2005)
12. Damoulas, T., Girolami, M.A.: Probabilistic multi-class multi-kernel learning: on protein fold recognition and remote homology detection. Bioinformatics **24**, 1264–1270 (2008)
13. Gönen, M.: Bayesian efficient multiple kernel learning. In: ICML, pp. 1–8 (2012)
14. Wang, H., Schmid, C.: Action recognition with improved trajectories. In: ICCV, pp. 3551–3558 (2013)
15. Perronnin, F., Dance, C.: Fisher kernels on visual vocabularies for image categorization. In: CVPR, pp. 1–8 (2007)
16. Beal, M.J.: Variational Algorithms for Approximate Bayesian Inference. University of London, London (2003)
17. Schuldt, C., Laptev, I., Caputo, B.: Recognizing human actions: a local SVM approach. In: ICPR, vol. 3, pp. 32–36 (2004)
18. Rodriguez, M.D., Ahmed, J., Shah, M.: Action mach a spatio-temporal maximum average correlation height filter for action recognition. In: CVPR, pp. 1–8 (2008)
19. Kuehne, H., Jhuang, H., Garrote, E., Poggio, T., Serre, T.: HMDB: a large video database for human motion recognition. In: ICCV, pp. 2556–2563 (2011)
20. Sheng, B., Yang, W., Sun, C.: Action recognition using direction-dependent feature pairs and non-negative low rank sparse model. Neurocomputing **158**, 73–80 (2015)
21. Zhang, H., Zhou, W., Reardon, C., Parker, L.E.: Simplex-based 3d spatio-temporal feature description for action recognition. In: CVPR, pp. 2067–2074 (2014)
22. Wu, J., Zhang, Y., Lin, W.: Towards good practices for action video encoding. In: CVPR, pp. 2577–2584 (2014)
23. Sun, L., Jia, K., Chan, T., Fang, Y., Wang, G., Yan, S.: Dl-sfa: deeply-learned slow feature analysis for action recognition. In: CVPR, pp. 2625–2632 (2014)
24. Yang, X., Tian, Y.L.: Action recognition using super sparse coding vector with spatio-temporal awareness. In: ECCV, pp. 727–741 (2014)
25. Veeriah, V., Zhuang, N., Qi, G.: Differential recurrent neural networks for action recognition. In: ICCV, pp. 4041–4049 (2015)
26. Lan, T., Zhu, Y., Zamir, A.R., Savarese, S.: Action recognition by hierarchical mid-level action elements. In: ICCV, pp. 4552–4560 (2015)
27. Shao, L., Liu, L., Yu, M.: Kernelized multiview projection for robust action recognition. IJCV 1–15 (2015)
28. Wang, D., Shao, Q., Li, X.: A new unsupervised model of action recognition. In: ICIP, pp. 1160–1164 (2015)

29. Liu, A.A., Su, Y.T., Nie, W.Z., Kankanhalli, M.: Hierarchical clustering multi-task learning for joint human action grouping and recognition. T-PAMI, 1–14 (2016)
30. Liu, L., Shao, L., Li, X., Lu, K.: Learning spatio-temporal representations for action recognition: a genetic programming approach. IEEE Trans. Cybern. **46**, 158–170 (2016)

Audio Visual Speech Recognition Using Deep Recurrent Neural Networks

Abhinav Thanda[✉] and Shankar M. Venkatesan

Samsung R&D Institute India, Bangalore, Bangalore, India
{abhinav.t89,s.venkatesan}@samsung.com

Abstract. In this work, we propose a training algorithm for an audio-visual automatic speech recognition (AV-ASR) system using deep recurrent neural network (RNN). First, we train a deep RNN acoustic model with a Connectionist Temporal Classification (CTC) objective function. The frame labels obtained from the acoustic model are then used to perform a non-linear dimensionality reduction of the visual features using a deep bottleneck network. Audio and visual features are fused and used to train a fusion RNN. The use of bottleneck features for visual modality helps the model to converge properly during training. Our system is evaluated on GRID corpus. Our results show that presence of visual modality gives significant improvement in character error rate (CER) at various levels of noise even when the model is trained without noisy data. We also provide a comparison of two fusion methods: feature fusion and decision fusion.

Keywords: Audio-visual speech recognition · Connectionist Temporal Classification · Recurrent neural network

1 Introduction

Audio-visual automatic speech recognition (AV-ASR) is a case of multi-modal analysis in which two modalities (audio and visual) complement each other to recognize speech. Incorporating visual features, such as speaker's lip movements and facial expressions, into automatic speech recognition (ASR) systems has been shown to improve their performances especially under noisy conditions. To this end several methods have been proposed which traditionally included variants of GMM/HMM models [3,5]. More recently AV-ASR methods based on deep neural networks (DNN) [14,21,23] have been proposed.

End-to-end speech recognition methods based on RNNs trained with CTC objective function [10,11,19] have come to the fore recently and have been shown to give performances comparable to that of DNN/HMM. The RNN trained with CTC directly learns a mapping between audio feature frames and character/phoneme sequences. This method eliminates the need for an intermediate step of training GMM/HMM model, thereby simplifying the training procedure. To our knowledge, so far AV-ASR systems based on RNN trained with CTC have not been explored.

© Springer International Publishing AG 2017
F. Schwenker and S. Scherer (Eds.): MPRSS 2016, LNAI 10183, pp. 98–109, 2017.
DOI: 10.1007/978-3-319-59259-6_9

In this work, we design and evaluate an audio-visual ASR (AV-ASR) system using deep recurrent neural network (RNN) and CTC objective function. The design of an AV-ASR system includes the tasks of visual feature engineering, and audio-visual information fusion. Figure 1 shows the AV-ASR pipeline at test time. This work mainly deals with the visual feature extraction and processing steps and training protocol for the fusion model. Proper visual features are important especially in the case of RNNs as RNNs are difficult to train. Bottleneck features used in tandem with audio features are known to improve ASR performance [7,12,28]. We employ a similar idea in order to improve the discriminatory power of video features. We show that this helps the RNN to converge properly when compared with raw DCT features. Finally, we compare the performances of feature fusion and decision fusion methods.

The paper is organized as follows: Sect. 2 presents the prior work on AV-ASR. Bi-directional RNN and its training using CTC objective function are discussed in Sect. 3. Section 4 describes the feature extraction steps for audio and visual modalities. In Sect. 5 different fusion models are explained. Section 6 explains the training protocols and experimental results. Finally, we summarize our work in Sect. 7.

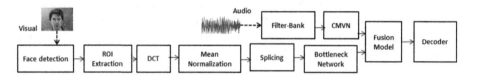

Fig. 1. Pipeline of AV-ASR system at test time. Fusion

2 Related Work

The differences between various AV-ASR systems lie chiefly in the methods employed for visual feature extraction and audio-visual information fusion. Visual feature extraction methods can be of 3 types [24]: 1. Appearance based features where each pixel in the mouth region of the speaker (ROI) is considered to be informative. Usually a transformation such as DCT or PCA is applied to the ROI to reduce the dimensions. Additional feature processing such as mean normalization, intra-frame and inter-frame LDA may be applied [15,24]. 2. Shape based features utilize the geometric features such as height, width and area of the lip region or build a statistical model of the lip contours whose parameters are used as features. 3. Combination of appearance and shape based features.

Fusion methods can be broadly divided into two types [16,24]: 1. Feature fusion 2. Decision fusion. Feature fusion models perform a low level integration of audio and visual features and this involves a single model which is trained on the fused features. Feature fusion may include a simple concatenation of features or feature weighting and is usually followed by a dimensionality reduction transformation like LDA. On the other hand, Decision fusion is applied in cases where the output classes for the two modalities are same. Various decision fusion methods based on variants of HMMs have been proposed [3,5]. In Multistream HMM the emission probability of a state of audio-visual system is obtained by a linear combination of log-likelihoods of individual streams for that state. The parameters of HMMs for individual streams can be estimated separately or jointly. While multistream HMM assumes state level synchrony between the two streams, some methods [2,3] such as coupled HMM [3] allow for asynchrony between two streams. For a detailed survey on HMM based AV-ASR systems we refer the readers to [16,24]

Application of deep learning to multi-modal analyses was presented in [22] which describes multi-modal, cross-modal and shared representation learning and their applications to AV-ASR. In [14], Deep Belief Networks (DBN) are explored. In [21] the authors train separate networks for audio and visual inputs and fuse the final layers of two networks, and then build a third DNN with the fused features. In addition, [21] presents a new DNN architecture with a bilinear soft-max layer which further improves the performance. In [23] a deep de-noising auto-encoder is used to learn noise robust speech features. The auto-encoder is trained with MFCC features of noisy speech as input and reconstructs clean features. The outputs of final layer of the auto-encoder are used as audio features. A CNN is trained with images from the mouth region as input and phoneme labels as output. The final layers of the two networks are then combined to train a multi-stream HMM.

3   Sequence Labeling Using RNN

The following notations are adopted in this paper. For an utterance u of length T_u, $\mathbf{O}_a^u = (\overline{O}_{a,1}^u, \overline{O}_{a,2}^u, ..., \overline{O}_{a,T_u}^u)$ and $\mathbf{O}_v^u = (\overline{O}_{v,1}^u, \overline{O}_{v,2}^u, ..., \overline{O}_{v,T_u}^u)$ denote the observation sequences of audio and visual frames where $\overline{O}_{a,t} \in \mathbb{R}^{d_a}$ and $\overline{O}_{v,t} \in \mathbb{R}^{d_v}$. We assume equal frame rates for audio and visual inputs which is ensured in experiments by means of interpolation. $\mathbf{O}_{av}^u = (\overline{O}_{av,1}^u, \overline{O}_{av,2}^u, ..., \overline{O}_{av,T_u}^u)$ where $\overline{O}_{av,t}^u = [\overline{O}_{a,t}^u, \overline{O}_{v,t}^u] \in \mathbb{R}^{d_{av}}$ where $d_{av} = d_a + d_v$ denotes the concatenated features at time t for utterance u. The corresponding label sequence is given by $l = (l_1, l_2, ..., l_{S_u})$ where $S_u \leq T_u$ and $l_i \in L$ where L is the set of English letters and an additional element representing a space. For ease of representation, we drop the utterance index u. All the models described in this paper are character based.

3.1   Bi-directional RNN

RNNs are a class of neural networks used to map sequences to sequences. This is possible because of the feedback connections between hidden nodes. In a bi-directional RNN, the hidden layer has two components each corresponding to forward (past) and backward (future) connections. For a given input sequence

$\mathbf{O} = (\overline{O}_1, \overline{O}_2, ..., \overline{O}_T)$, the output of the network is calculated as follows: forward pass through forward component of the hidden layer at a given instant t is given by

$$\overline{h}_t^f = g(\mathbf{W}_{ho}^f \overline{O}_t + \mathbf{W}_{hh}^f \overline{h}_{t-1}^f + \overline{b}_h^f) \tag{1}$$

where $\mathbf{W}_{ho}^f$ is the input-to-hidden weights for forward component, $\mathbf{W}_{hh}^f$ corresponds to hidden-to-hidden weights between forward components, and $\overline{b}_h^f$ is the forward component bias. g is a non-linearity depending on the choice of the hidden layer unit. Similarly, forward pass through the backward component of the hidden layer is given by

$$\overline{h}_t^b = g(\mathbf{W}_{ho}^b \overline{O}_t + \mathbf{W}_{hh}^b \overline{h}_{t-1}^b + \overline{b}_h^b) \tag{2}$$

where $\mathbf{W}_{ho}^b$, $\mathbf{W}_{hh}^b$, $\overline{b}_h^b$ are the corresponding parameters for the backward component. The input to next layer is the concatenated vector $[\mathbf{h}_t^f, \mathbf{h}_t^b]$. In a deep RNN multiple such bidirectional hidden layers are stacked.

RNNs are trained using Back-Propagation Through Time (BPTT) algorithm. The training algorithm suffers from vanishing gradients problem which is overcome by using a special unit in hidden layer called the Long Short Term Memory (LSTM) [8,13].

3.2 Connectionist Temporal Classification

DNNs used in ASR systems are frame-level classifiers i.e., each frame of the input sequence requires a class label in order for the DNN to be trained. The frame-level labels are usually HMM states, obtained by first training a GMM/HMM model and then by forced alignment of input sequences to the HMM states. CTC objective function [9,10] obviates the need for such alignments as it enables the network to learn over all possible alignments.

Let the input sequence be $\mathbf{O} = (\overline{O}_1, \overline{O}_2, ..., \overline{O}_T)$ and a corresponding label sequence $\mathbf{l} = (l_1, l_2, ..., l_S)$ where $S \leq T$. The RNN employs a soft-max output layer containing one node for each element in L' where $L' = L \cup \{\phi\}$. The number of output units is $|L'| = |L| + 1$. The additional symbol ϕ represents a blank label meaning that the network has not produced an output for that input frame. The additional blank label at the output allows us to define an alignment π of length T containing elements of L'. For example, $(A\phi\phi M\phi), (\phi A\phi\phi M)$ are both alignments of length 5 for the label sequence AM. Accordingly, a many to one map $B : L'^T \longmapsto L^{\leq T}$ can be defined which generates the label sequence from an alignment.

Assuming that the posterior probabilities obtained at soft-max layer, at each instant are independent we get

$$P(\pi|\mathbf{O}) = \prod_{t=1}^{T} P(k_t|\overline{O}_t) \tag{3}$$

where $k \in L'$ and

$$P(k_t|\overline{O}_t) = \frac{\exp(y_t^k)}{\Sigma_{k'} \exp(y_t^{k'})} \tag{4}$$

where y_t^k is the input to node k of the soft-max layer at time t

The likelihood of the label sequence given an observation sequence can be calculated by summing (3) over all possible alignments.

$$P(\mathbf{l}|\mathbf{O}) = \sum_{\pi \in B^{-1}(\mathbf{l})} P(\pi|\mathbf{O}) \tag{5}$$

The goal is to maximize the log-likelihood $\log P(\mathbf{l}|\mathbf{O})$ estimation of a label sequence given an observation sequence. Equation 5 is computationally intractable since the number of alignments increases exponentially with the number of labels. For efficient computation of (5), forward-backward algorithm is used.

4   Feature Extraction

4.1   Audio Features

The sampling rate of audio data is converted to 16 kHz. For each frame of speech signal of 25 ms duration, filter-bank features of 40 dimensions are extracted. The filter-bank features are mean normalized and Δ and $\Delta\Delta$ features are appended. The final 120 dimensional features are used as audio features.

4.2   Visual Features

The video frame rate is increased to match the rate of audio frames through interpolation. For AV-ASR, the ROI for visual features is the region surrounding the speaker's mouth. Each frame is converted to gray scale and face detection is performed using Viola-Jones algorithm. The 64×64 lip region is extracted by detecting 68 landmark points [17] on the speakers face, and cropping the ROI surrounding speakers mouth and chin. 100 dimensional DCT features are extracted from the ROI.

After several experiments of training with DCT features, we found that RNN training either exploded or converged poorly. In order improve the discriminatory power of the visual features, we perform non-linear dimensionality reduction of the features using a deep bottleneck network. Bottleneck features are obtained by training a neural network in which one of the hidden layers has relatively small dimension. The DNN is trained using cross-entropy cost function with character labels as output. The frame-level character labels required for training the DNN are obtained by first training an acoustic model (RNN_a) and then obtaining the outputs from the final soft-max layer of RNN_a.

The DNN configuration is given by $dim - 1024 - 1024 - 40 - 1024 - opdim$ where $dim = 1100$ and is obtained by splicing each 100 dimensional video frame

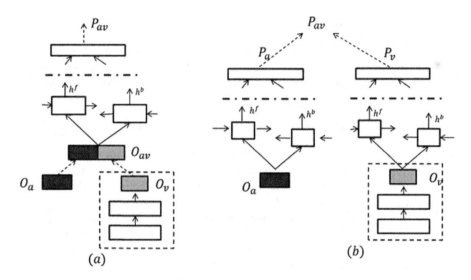

Fig. 2. Fusion models (a) Feature fusion (b) Decision fusion. The bottleneck network for visual feature extraction is enclosed in the dotted box.

with a context of 10 frames - 5 on each side. $opdim = |L'|$. After training, the last 2 layers are discarded and 40-dimensional outputs are used as visual features. The final dimension of visual feature vector is 120 including the Δ and $\Delta\Delta$ features.

5 Fusion Models

In this work, the fusion models are character based RNNs trained using CTC objective function i.e. L' is the set of English alphabet including a blank label. The two fusion models are shown in Fig. 2.

5.1 Feature Fusion

In feature fusion technique, a single RNN_{av} is trained by concatenating the audio and visual features using the CTC objective function. In the test phase, at each instant the concatenated features are forward propagated through the network. In the CTC decoding step, the posterior probabilities obtained at the soft-max layer are converted to pseudo log-likelihoods [26] as

$$\log P_{av}(\overline{O}_{av,t}|k) = \log P_{av}(k|\overline{O}_{av,t}) - \log P(k) \qquad (6)$$

where $k \in L'$ and $P(k)$ is the prior probability of class k obtained from the training data [19].

5.2   Decision Fusion

In decision fusion technique the audio and visual modalities are modeled by separate networks, RNN_a and RNN_v respectively. RNN_v is a lip-reading system. The networks are trained separately. In the test phase, for a given utterance the frame level, the pseudo log-likelihoods of RNN_a and RNN_v are combined as

$$\log P_{av}(\overline{O}_{a,t}, \overline{O}_{v,t}|k) = \gamma \log P_a(k|\overline{O}_{a,t}) + (1 - \gamma) \log P_v(k|\overline{O}_{v,t}) - \log P(k) \quad (7)$$

where $0 \leq \gamma \leq 1$ is a parameter dependent on the noise level and the reliability of each modality [5]. For example, at higher levels of noise in audio input, a low value of γ is preferred. In this work, we adapt the parameter γ for each utterance based on KL-divergence measure between the posterior probability distributions of RNN_a and RNN_v. The divergence between the posterior probability distributions is expected to vary as the noise in the audio modality increases. The KL-divergence is scaled to a value in $[0, 1]$ using logistic sigmoid. The parameter b was determined empirically from validation dataset.

$$D_{KL}(P_v||P_a) = \sum_i P_v log P_a \quad (8)$$

where we consider the posteriors of RNN_v as the true distribution based on the assumption that video input is always free from noise.

$$\gamma = \frac{1}{1 + exp(-D_{KL} + b)} \quad (9)$$

6   Experiments

The system was trained and tested on GRID audio-visual corpus [4]. GRID corpus is a collection of audio and video recordings of 34 speakers (18 male, 16 female) each uttering a 1000 sentences. Each utterance has a fixed length of approximately 3 s. The total number of words in the vocabulary is 51. The syntactic structures of all sentences are similar as shown below.

$< command >$ $< color >$ $< preposition >$ $< letter >$ $< digit >$ $< adverb >$
Ex. PLACE RED AT M ZERO PLEASE

6.1   Training

In the corpus obtained, the video recordings for speaker 21 were not available. In addition, 308 utterances by various speakers could not be processed due to various errors. The dataset in effect consisted of 32692 utterances 90% of the which (containing 29423 utterances) was used for training and cross validation while the remaining (10%) data was used as test set. Both training and test data contain utterances from all of the speakers. Models were trained and tested using Kaldi speech recognition tool kit [25], Kaldi+PDNN [18] and EESEN framework [19].

RNN$_a$ **Acoustic Model.** RNN_a contains 2 bi-directional LSTM hidden layers. Input to the network is 120-dimensional vector containing filter-bank coefficients along with Δ and $\Delta\Delta$ features. The model parameters are randomly initialized within the range $[-0.1, 0.1]$. The initial learning rate is set to 0.00004. Learning rate adaption is performed as follows: when the improvement in accuracy on the cross-validation set between two successive epochs falls below 0.5%, the learning rate is halved. The halving continues for each subsequent epoch until the training stops when the increase in frame level accuracy is less than 0.1%.

Deep Bottleneck Network. The training protocol similar to [26] was followed to train the bottleneck network. Input video features are mean normalized and spliced. Cross-entropy loss function is minimized using mini-batch Stochastic Gradient Descent (SGD). The frames are shuffled randomly before each epoch. Batch size is set to 256 and initial learning rate is set to 0.008. Learning rate adaptation similar to acoustic model is employed.

RNN$_v$**-Lip Reader.** RNN_v is trained with bottleneck network features as input. The network architecture and training procedure is same as RNN_a. Figure 3 depicts the learning curves when trained with bottleneck features and DCT features. The figure shows that bottleneck features are helpful in proper convergence of the model.

RNN$_{av}$. The feature fusion model RNN_{av} consists of 3 bi-directional LSTM hidden layers. The input dimension is 240, corresponding to filter-bank coefficients of audio modality, bottleneck features of visual modality and their respective Δ features. The initialization and learning rate adaption are similar to acoustic model training. However, the learning rate adaptation is employed only after a minimum number of (in this case 20) epochs are completed.

During each utterance in an epoch we first present the fused audio-visual fused input sequence followed by the input sequence with audio input set to very low values. This prevents the RNN_{av} from over-fitting to audio only inputs. Thus the effective number of sequences presented to the network in a given epoch is twice the total number of training utterances (AV and V features). After the training with AV and V features we train the network once again with two epochs of audio only utterances obtained by turning off the visual modality.

6.2   Results

The audio-visual model is tested with three levels of babble noise 0 dB SNR, 10 dB SNR and clean audio. Noise was added to test data artificially by mixing babble noise with clean audio .wav files. In order to show the importance of visual modality under noisy environment, the model is tested with either audio or video inputs turned off. A token WFST [19] is used to map the paths to their corresponding label sequences. The token WFST obtains this mapping by removing all the blanks and repeated labels. Character Error Rate (CER) is obtained from the decoded and expected label sequences by calculating the edit distance between them. The CER results are shown in Table 1.

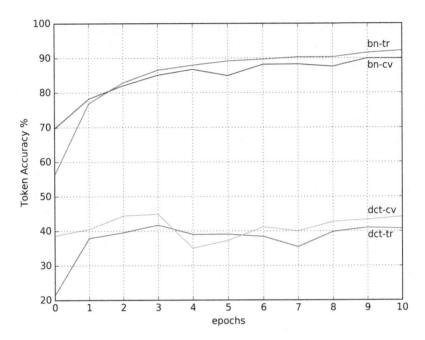

Fig. 3. Learning curves for bottleneck (bn) features and DCT features for training (tr) and validation (cv) data sets.

We observe that with clean audio input, audio only RNN_a performs significantly better (CER 2.45%) compared to audio-visual RNN_{av} (CER 5.74%). However as audio becomes noisy, the performance of RNN_a deteriorates significantly whereas the performance of RNN_{av} remains relatively stable. Under noisy conditions the feature fusion model behaves as if it is not receiving any input from the audio modality.

Table 1 also gives a comparison between feature fusion model and decision fusion model. We find that feature fusion model performs better than decision fusion model in all cases except under clean audio conditions. The poor CER of RNN_a, RNN_v model indicates that the frame level predictions between RNN_a and RNN_v are not synchronous. However, both the fusion models provide significant gains under noisy audio inputs. While there is large difference between RNN_a and other models with clean inputs, we believe this difference is due to the nature of dataset and will reduce with larger datasets.

Comparison with Lip-Reading Systems. While a number of AV-ASR models exist, to our knowledge none of the methods were trained and tested on GRID corpus. However, results on several lip-reading systems (visual only inputs) on GRID corpus have been reported. Table 2 gives a comparison of lip-reading systems which employ recurrent neural networks. LipNet is a recent independent work which uses spatio-temporal convolutions and Gated Recurrent Units. It is

Table 1. % CER comparison for feature fusion (RNN_{av}) and decision fusion (RNN_a, RNN_v) models. RNN_a is the acoustic model and RNN_v is the lip reader.

Feature fusion				Decision fusion				
Model	Input		CER %	Model	Input			CER %
	Audio	Visual			Audio	Visual		
RNN_{av}	Clean	OFF	7.35	RNN_a, RNN_v	Clean	OFF		2.45
RNN_{av}	Clean	ON	5.74	RNN_a, RNN_v	Clean	ON		8.46
RNN_{av}	OFF	ON	11.42	RNN_a, RNN_v	OFF	ON		11.06
RNN_{av}	10 SNR dB	OFF	38.31	RNN_a, RNN_v	10 SNR dB	OFF		23.83
RNN_{av}	10 SNR dB	ON	10.24	RNN_a, RNN_v	10 SNR dB	ON		14.83
RNN_{av}	0 SNR dB	OFF	59.65	RNN_a, RNN_v	0 SNR dB	OFF		59.27
RNN_{av}	0 SNR dB	ON	11.57	RNN_a, RNN_v	0 SNR dB	ON		16.84

trained using CTC at sentence level like our model whereas the RNN-LSTM model in [27] is trained at word level. However, in contrast to LipNet our aim in this paper was to present a noise-robust ASR which utilizes both audio and visual modalities which we believe will perform better with larger vocabulary datasets. Our model has the potential to switch from audio to a mixed modality (by turning the camera on) based on an SNR measure (where we define the signal as a continually discernible linguistic content from an utterance as measured perhaps using KL divergence described before). The %CER for LipNet [1] and the RNN-LSTM model of Wand et al., [27] are reported from [1].

Table 2. % CER comparison of lip-reading systems employing RNNs. The audio modality for the model in the last row is turned off.

Method	CER %
LipNet	1.90
Wand et al.	15.20
RNN_v	11.06
RNN_{av}	11.42

7   Conclusions and Future Work

In this work we presented an audio-visual ASR system using deep RNNs trained with CTC objective function. We described a feature processing step for visual features using deep bottleneck layer and showed that it helps in faster convergence of RNN model during training. We presented a training protocol in which either of the modalities is turned off during training in order to avoid dependency on a single modality. Our results indicate that the trained model is robust to noise. In addition, we compared fusion strategies at the feature level and at the decision level.

While the use of bottleneck features for visual modality helps in training, it requires frame level labels which involves an additional step of training audio RNN. Therefore, our system is not yet end-to-end. Our experiments in visual feature engineering with unsupervised methods like multi-modal auto-encoder [22] did not produce remarkable results. Currently, we are exploring visual features like curl and divergence of optical flow field using the Fourier Transform based on Clifford Algebra [6,20]. In future work we intend to explore other unsupervised methods for visual feature extraction such as canonical correlation analysis.

References

1. Assael, Y.M., Shillingford, B., Whiteson, S., de Freitas, N.: Lipnet: Sentence-level lipreading. arXiv preprint arXiv:1611.01599 (2016)
2. Bengio, S.: Multimodal speech processing using asynchronous hidden markov models. Inform. Fusion **5**(2), 81–89 (2004)
3. Brand, M., Oliver, N., Pentland, A.: Coupled hidden markov models for complex action recognition. In: Proceedings IEEE Computer Society Conference on Computer vision and pattern recognition, pp. 994–999. IEEE (1997)
4. Cooke, M., Barker, J., Cunningham, S., Shao, X.: An audio-visual corpus for speech perception and automatic speech recognition. J. Acoust. Soc. Am. **120**(5), 2421–2424 (2006)
5. Dupont, S., Luettin, J.: Audio-visual speech modeling for continuous speech recognition. IEEE Trans. Multimedia **2**(3), 141–151 (2000)
6. Ebling, J., Scheuermann, G.: Clifford fourier transform on vector fields. IEEE Trans. Vis. Comput. Graph. **11**(4), 469–479 (2005)
7. Gehring, J., Miao, Y., Metze, F., Waibel, A.: Extracting deep bottleneck features using stacked auto-encoders. In: 2013 IEEE International Conference on Acoustics, Speech and Signal Processing, pp. 3377–3381. IEEE (2013)
8. Graves, A.: Supervised Sequence Labelling with Recurrent Neural Networks. SCI, vol. 385, pp. 15–35. Springer, Heidelberg (2012)
9. Graves, A., Fernández, S., Gomez, F., Schmidhuber, J.: Connectionist temporal classification: labelling unsegmented sequence data with recurrent neural networks. In: Proceedings of the 23rd International Conference on Machine Learning, pp. 369–376. ACM (2006)
10. Graves, A., Jaitly, N.: Towards end-to-end speech recognition with recurrent neural networks. In: ICML, vol. 14, pp. 1764–1772 (2014)
11. Hannun, A., Case, C., Casper, J., Catanzaro, B., Diamos, G., Elsen, E., Prenger, R., Satheesh, S., Sengupta, S., Coates, A., et al.: Deep speech: scaling up end-to-end speech recognition. arXiv preprint arXiv:1412.5567 (2014)
12. Hermansky, H., Ellis, D.P., Sharma, S.: Tandem connectionist feature extraction for conventional hmm systems. In: Proceedings of the 2000 IEEE International Conference on Acoustics, Speech, and Signal Processing, ICASSP 2000, vol. 3, pp. 1635–1638. IEEE (2000)
13. Hochreiter, S., Schmidhuber, J.: Long short-term memory. Neural Comput. **9**(8), 1735–1780 (1997)
14. Huang, J., Kingsbury, B.: Audio-visual deep learning for noise robust speech recognition. In: 2013 IEEE International Conference on Acoustics, Speech and Signal Processing, pp. 7596–7599. IEEE (2013)

15. Huang, J., Potamianos, G., Neti, C.: Improving audio-visual speech recognition with an infrared headset. In: AVSP 2003-International Conference on Audio-Visual Speech Processing (2003)
16. Katsaggelos, A.K., Bahaadini, S., Molina, R.: Audiovisual fusion: challenges and new approaches. Proc. IEEE **103**(9), 1635–1653 (2015)
17. Kazemi, V., Sullivan, J.: One millisecond face alignment with an ensemble of regression trees. In: Proceedings of the IEEE Conference on Computer Vision and Pattern Recognition, pp. 1867–1874 (2014)
18. Miao, Y.: Kaldi+pdnn: building dnn-based asr systems with kaldi and pdnn. arXiv preprint arXiv:1401.6984 (2014)
19. Miao, Y., Gowayyed, M., Metze, F.: Eesen: End-to-end speech recognition using deep rnn models and wfst-based decoding. In: 2015 IEEE Workshop on Automatic Speech Recognition and Understanding (ASRU), pp. 167–174. IEEE (2015)
20. Mohammadzade, H., Bruton, L.T.: A simultaneous div-curl 2D clifford fourier transform filter for enhancing vortices, sinks and sources in sampled 2D vector field images. In: IEEE International Symposium on Circuits and Systems, ISCAS 2007, pp. 821–824. IEEE (2007)
21. Mroueh, Y., Marcheret, E., Goel, V.: Deep multimodal learning for audio-visual speech recognition. In: 2015 IEEE International Conference on Acoustics, Speech and Signal Processing (ICASSP), pp. 2130–2134. IEEE (2015)
22. Ngiam, J., Khosla, A., Kim, M., Nam, J., Lee, H., Ng, A.Y.: Multimodal deep learning. In: Proceedings of the 28th International Conference on Machine Learning (ICML 2011), pp. 689–696 (2011)
23. Noda, K., Yamaguchi, Y., Nakadai, K., Okuno, H.G., Ogata, T.: Audio-visual speech recognition using deep learning. Appl. Intell. **42**(4), 722–737 (2015)
24. Potamianos, G., Neti, C., Gravier, G., Garg, A., Senior, A.W.: Recent advances in the automatic recognition of audiovisual speech. Proc. IEEE **91**(9), 1306–1326 (2003)
25. Povey, D., Ghoshal, A., Boulianne, G., Burget, L., Glembek, O., Goel, N., Hannemann, M., Motlicek, P., Qian, Y., Schwarz, P., et al.: The kaldi speech recognition toolkit. In: IEEE 2011 Workshop on Automatic Speech Recognition and Understanding. No. EPFL-CONF-192584. IEEE Signal Processing Society (2011)
26. Veselý, K., Ghoshal, A., Burget, L., Povey, D.: Sequence-discriminative training of deep neural networks. In: INTERSPEECH, pp. 2345–2349 (2013)
27. Wand, M., Koutník, J., Schmidhuber, J.: Lipreading with long short-term memory. In: 2016 IEEE International Conference on Acoustics, Speech and Signal Processing (ICASSP), pp. 6115–6119. IEEE (2016)
28. Yu, D., Seltzer, M.L.: Improved bottleneck features using pretrained deep neural networks. In: Interspeech, vol. 237, p. 240 (2011)

Audio-Visual Recognition of Pain Intensity

Patrick Thiam[1]([✉]), Viktor Kessler[1], Steffen Walter[2], Günther Palm[1],
and Friedhelm Schwenker[1]

[1] Institute of Neural Information Processing, Ulm University,
James-Franck-Ring, 89061 Ulm, Germany
{patrick.thiam,viktor.kessler,steffen.walter,
guenther.palm,friedhelm.schwenker}@uni-ulm.de
[2] Department of Psychosomatic Medicine and Psychotherapy,
Ulm University, Ulm, Germany

Abstract. In this work, a multi-modal pain intensity recognition system based on both audio and video channels is presented. The system is assessed on a newly recorded dataset consisting of several individuals, each subjected to 3 gradually increasing levels of painful heat stimuli under controlled conditions. The assessment of the dataset consists of the extraction of a multitude of features from each modality, followed by an evaluation of the discriminative power of each extracted feature set. Finally, several fusion architectures, involving early and late fusion, are assessed. The temporal availability of the audio channel is taken in consideration during the assessment of the fusion architectures.

Keywords: Pain intensity recognition · Decision fusion · Multi-modal affect recognition · Random Forests

1 Introduction

An unreliable and inconsistent assessment of pain might lead to an unsuitable and insufficient therapy. Such a scenario might occur due to countless factors, among others, old age, mental impairment or degenerative diseases. Consequently, instead of experiencing some relief, the patient would further suffer from physical impairment and psychological discomfort. A reliable and automatic pain recognition system would be beneficial, since it would allow a better assessment of pain intensity, thus a better choice of therapy that would considerably improve the quality of life of the patients.

In the last decades, approaches for automatic pain recognition have gone from uni-modal systems focusing on one unique and specific modality such as video signals [8,17,24] or bio-physiological signals [1,6,9,14], to multi-modal systems where several modalities are combined to improve the pain intensity recognition rate by using complementary features extracted from each of the modalities [2,15,16,22,25]. The most common modalities involved in the assessment of the pain intensity include both video and bio-physiological channels. To the author's

© Springer International Publishing AG 2017
F. Schwenker and S. Scherer (Eds.): MPRSS 2016, LNAI 10183, pp. 110–126, 2017.
DOI: 10.1007/978-3-319-59259-6_10

knowledge there have not been any studies involving the audio channel as an additional modality for the assessment of pain.

Therefore, the following work aims to investigate the applicability of the audio channel as well as the fusion of both audio and video channels in both participant dependent and independent pain recognition scenarios. Furthermore, a newly recorded dataset, upon which the present work is built, is presented. The dataset targets the assessment and evaluation of pain in a controlled environment, as well as the analysis of the influence of emotions on pain perception.

The remainder of this work is organised as follows. Section 2 consists of the description of the dataset. In Sect. 3 a description of the audio channel processing, feature extraction and assessment pipeline is provided. In Sect. 4 the video channel processing, feature extraction and assessment pipeline is described. The conducted fusion experiments as well as the corresponding results are presented in Sect. 5 followed by the discussion and conclusion in Sect. 6.

2   Dataset Description

The data utilized in the present work was recently collected with the goal of generating a multimodal corpus designed specifically for research in the domain of emotion and pain recognition. It consists of 40 participants (20 male, 20 female), each subjected to two sessions of experiments of about 40 min each, during which several pain and emotion stimuli were triggered and the demeanour of each participant was recorded using audio, video and bio-physiological sensors.

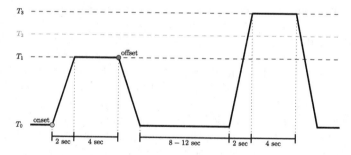

Fig. 1. Pain stimulation. T_0: baseline temperature (32 °C); T_1: pain threshold temperature; T_2: intermediate temperature; T_3: pain tolerance temperature. (Color figure online)

The pain stimuli were elicited through heat generated by a Medoc Pathway thermal simulator[1]. The experiment was repeated for each participant twice, each time with the ATS thermode attached to a different forearm (left and right). Before the data was recorded, each participant's pain threshold temperature and pain tolerance temperature were determined. Based on both temperatures, an intermediate heat stimulation temperature was computed such that

---

[1] http://medoc-web.com/products/pathway-model-ats/.

the range between both the threshold and tolerance temperatures was divided into 2 equally spaced ranges.

A specific emotional elicitation was triggered simultaneously to each pain elicitation in the form of pictures and video clips. The latter were carefully selected with the purpose of triggering specific emotional responses. This allowed a categorisation of the emotion stimuli using a two dimensional valence-arousal space in the following groups: positive (positive valence, high arousal); negative (negative valence, low arousal); neutral (neutral valence, neutral arousal).

Each heat temperature (pain stimulation) was triggered randomly 30 times with a randomised pause lasting between 8 and 12 s between consecutive stimuli. The randomised and simultaneous emotion stimuli were distributed for each heat temperature (pain stimulation) as well as the baseline temperature (no pain stimulation) as follows: 10 positive, 10 negative and 10 neutral emotion elicitations. Each stimulation consisted of a 2 s onset during which the temperature was gradually elevated starting from the baseline temperature until the specific heat temperature was reached. Following, the attained temperature was maintained for 4 s before being gradually dropped until the baseline temperature was reached. A recovery phase of 8–12 s followed before the next pain stimulation was elicited (see Fig. 1 for more details).

Therefore, each participant is represented by two sets of data, each one representing the experiments conducted on each forearm (left and right). Each dataset consists of 120 pain stimuli with 30 stimuli pro temperature (T_0: baseline, T_1: threshold, T_2: intermediate, T_3: tolerance), and 120 emotion stimuli with 40 stimuli pro emotion category (positive, negative, neutral).

The synchronous data recorded from the experiments consists of 3 high resolution video streams from 3 different perspectives, 2 audio lines recorded respectively from a headset and a directional microphone, and 4 physiological channels, namely the electromyographic activity of the trapezius muscle (EMG), the galvanic skin response (GSR), the electrocardiogram (ECG) and the respiration (RSP). Furthermore, an additional video and audio stream were recorded using the Microsoft Kinect sensor.

The focus of the present work is the investigation of the relevance of both audio and video channels regarding the task of pain intensity recognition. Thus the recognition of the different categories of emotion or the impact of the emotion stimuli on pain recognition will not be investigated.

3   Audio Channel Assessment

Since the conducted experiments did not include any type of verbal interaction, the recorded audio signals consist mostly of breathing noises and sporadic moaning sounds. These sounds represent the unique material that has to be exploited in order to discriminate between the different pain stimulation levels. Hence, the current work is based uniquely on the signals recorded with the headset which are more suitable because of the headset's proximity to the nasolabial region. The audio recordings from the Microsoft Kinect sensor as well as those from the directional microphone could not capture the breathing noises satisfactorily, thus were not further analysed. A preliminary analysis of the recordings was undertaken in order to define an optimal window within which the feature extraction and classification tasks should be realised.

3.1   Temporal Window Analysis

In order to determine an appropriate temporal window within which adequate features for the discrimination of the different pain stimuli could be extracted, an initial experiment, consisting of the analysis of the correlation between the intensity of a pain stimulation and the energy within the recorded audio signal, was undertaken. Therefore, the Root Mean Square (RMS) energy of each audio signal was computed from frames of 25 ms sampled at a rate of 10 ms. The focus was put on the stimulation phases during which the participants were subjected to the highest heat temperatures (T_3). From those phases, a 10 s window starting from the point when the stimulation temperature starts to increase (see onset in Fig. 1) was empirically chosen for further analysis after observing that most of the participants reacted to the stimuli within this window. Furthermore, the extracted RMS energy signals were preprocessed by first applying a Butterworth bandpass filter to get rid of out of range noises and subsequently applying a median filter in order to smooth the signals. Following the preprocessing of the signals, the median value of each single frame was computed over the entire 40 participants. The results can be seen in Fig. 2.

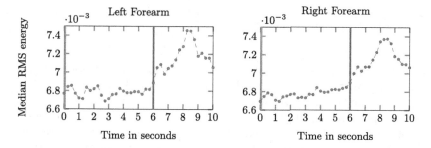

Fig. 2. Median RMS energy over 40 participants subjected to the highest heat temperature (T_3). Left: left forearm. **Right:** right forearm. The green plot represents the onset while the red plot represents the offset (see Fig. 1). (Color figure online)

The green plot corresponds to the onset and the red plot corresponds to the point when the heat temperature starts to sink (see offset in Fig. 1). The plotted data suggests that there is a considerable increase of energy following the heat stimulation. This increase of energy reaches a peak 2 or 3 s following the offset before decreasing gradually. This observation is in concordance with the observed demeanour of the participants during the experiments. Most of the participants would hold their breath as soon as the stimuli would get painful and would heavily breathe out as soon as the temperatures would start sinking, before breathing normally again.

Furthermore, in order to support this assumption, the data from both experiments (left forearm and right forearm) for all 40 participants was merged. The RMS energy was extracted and preprocessed as previously described from several windows of length $\pm l$ seconds ($l \in \{1, 2, 3, 4\}$) corresponding respectively to l seconds before ($-l$) and following ($+l$) the offset. The extracted energy was then summed for each participant and for each specific window before the median value was computed over all 40 participants. Moreover, the Wilcoxon signed rank test with two significance levels of 10^{-2} and 10^{-4} was computed for the significance assessment. The results are depicted in Fig. 3.

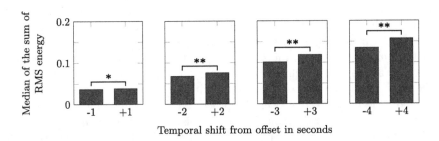

Fig. 3. Median of the sum of RMS energy over 40 participants subjected to the highest heat stimuli. The energy from both experiments (left forearm and right forearm) are merged together in order to generate the depicted results. The median energy is significantly higher a couple of seconds following the offset than preceding the offset with (*) $p < 10^{-2}$ and (**) $p \ll 10^{-4}$.

As assumed, Fig. 3 shows that the energy is significantly higher a couple of seconds following the offset than before the offset. In other words, the level of energy of the audio signal is low during the pain elicitation, before picking up within the phase during which the elicited temperature decreases. Therefore, the most relevant data for the audio-based classification task is not captured within the pain elicitation phase but rather some couple of seconds following the pain elicitation.

Subsequently an additional experiment was conducted to corroborate the findings of the previous experiments. A grid search was performed within the predefined 10 s window in order to determine an appropriate segment within which the best discrimination between the baseline temperature and the highest stimulation temperature could be attained. Thus, several segments, with lengths ranging from 4 to 6.5 s were defined for the feature extraction and subsequent classification. Those windows where temporally shifted starting from the onset. The temporal shifts ranged between 0 and 6 s.

Within each segment a simple unsupervised and threshold based voice activity detection algorithm was applied on the extracted RMS energy signal of each participant to distinguish between silence (or noise) frames and voice active frames. From the detected voice active frames 13 Mel frequency cepstral coefficients (MFCC) [11], each combined with its first and second order frame to frame difference, were extracted. Finally, the extracted features were used to perform a 10-fold cross validation participant dependent classification using a Random Forest classifier [5] with 30 decision trees. The model was trained using the features extracted at the frame level from the voice active segments of each window in the training set and subsequently applied on the unseen windows of the left out set. The model would assign a label to each frame of the unseen window before a simple majority vote would be applied in order to decide about the final label of the whole window.

The results of the classification are depicted in Fig. 4. The latter depicts the median of the classification accuracy of the baseline temperature against the highest heat stimulation temperature corresponding to the highest pain level, using the extracted MFCC features over the 40 participants, from each defined segment and for each forearm. The depicted results confirm the findings of the previous experiments, since the best classification performances are achieved for both forearms with a temporal shift from the onset between 4 and 6 s and a window length between 4 and 5.5 s. For the next experiments (including the fusion experiments) a window length of 4.5 s with a temporal shift from the onset of 4 s was selected for the audio channel.

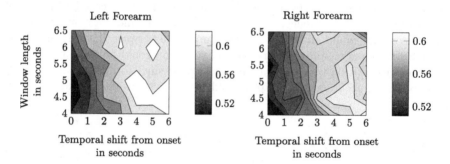

Fig. 4. Audio signal window assessment (Baseline temperature (T_0) vs Highest heat temperature (T_3)). Participant dependent 10-fold cross validation (median). The best classification performance is achieved with a temporal shift from the onset between 4 and 6 s and a window length between 4 and 5.5 s for both forearms.

3.2   Audio Features Extraction and Assessment

Based on the results described previously, several audio features were extracted from the voice active frames within the previously specified window, obtained by applying the threshold based voice activity detector. In addition to the MFCC features extracted in the previous phase, 5 Relative Spectral Perceptual Linear Predictive coefficients (RASTA-PLP) [10], each with its first and second order frame difference, as well as 8 Linear Predictive Coding Coefficients (LPC) [18] were extracted. Additionally, 14 spectral features (e.g. Hammarberg index, spectral flux, spectral centroid) were extracted as well as a combination of statistical features extracted from the zero crossing rate signal (ZCR), both the RMS- and log-energy, the voicing probability and the loudness contour [19]. All features were extracted from 25 ms frames, sampled at a rate of 10 ms and using the openSMILE features extraction tool [7].

Subsequently, the features were assessed by proceeding with a 10-fold participant dependent cross validation classification. A Random Forest model was trained as described in the previous Section and the label of each window was determined by majority voting as well. This assessment was done by using the data specific to both the baseline temperature (T_0) and the highest heat temperature (T_3). In other words, the assessment of the features was realised by performing a "pain" against "no pain" classification. Each set of features was first considered individually, followed by an early fusion of all extracted features. The classification results are depicted in Fig. 5. The depicted results show that the performance of each feature set is quite similar. No feature set significantly outperforms any other, and the mean classification accuracy of all feature sets including the early fusion set is located between 60% and 65% in both experiments (left and right forearm). Moreover the variance across the 40 participants is quite large. This can be explained by the fact that some of the participants are either unresponsive to the heat stimuli or no breathing noise could be recorded to a satisfactory extent. Hence, the recorded audio signal did not contain enough information that would allow a better discrimination between the different levels of pain. Still, some recordings were good enough and could be exploited to perform the classification task to a satisfactory extent since for at least 25% of the participants an accuracy above 70% could be achieved using the early feature fusion.

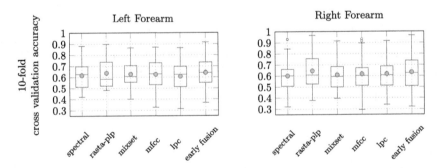

Fig. 5. Feature performance analysis (Baseline temperature (T_0) vs Highest heat temperature (T_3)). Participant dependent 10-fold cross validation. Within each box plot the mean of the classification accuracy across all 40 participants is depicted as a gray dot and the median as a red line. (Color figure online)

4   Video Channel Assessment

The video channel assessment is performed using only the video stream captured by the frontal camera (see Fig. 6) and consists of analysing the facial expressions of the participants during the experiments. In order to determine an optimal window within which the discrimination between the different pain intensities based uniquely on the analysis of the facial region can be achieved at a satisfactory extent, a set of facial landmarks (see Fig. 6(a)) is automatically detected and tracked using the facial behaviour analysis toolkit OpenFace [3].

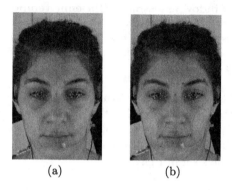

(a)   (b)

Fig. 6. Geometric features. (a) Facial landmarks. (b) Distances computed between the tracked facial landmarks.

Subsequently a set of 2D distances (see Fig. 6(b)) is computed between the tracked landmarks in order to capture the deformation of the facial area at the frame level. Throughout a defined temporal window, each of these distances yields a signal. These signals are low-pass filtered and the first and second derivatives of the filtered signals are computed. Several functionals (mean, median, maximum, minimum, range, standard deviation, kurtosis, skewness, first and second quartile, inter quartile, 1%-percentile, 99%-percentile, range of 1%-percentile and 99%-percentile) are subsequently applied on these signals to extract several statistical parameters that are used as geometric-based facial expression features for the classification task.

These features are extracted for each of the windows defined in Sect. 3.1 in order to perform a grid search. A participant specific 10-fold cross validation classification is subsequently performed using a Random Forest classifier with 300 decision trees and the median classification accuracy over all 40 participants is plotted for each defined window. The grid search is performed using the data specific to the baseline temperature (T_0) and the highest heat temperature (T_3). Figure 7 depicts the results of the performed grid search. For both datasets (left and right forearm) the best performance is yielded by choosing a window length between 5 and 6.5 s, with the corresponding temporal shift. This implies that the most relevant information for the classification task is located between the point

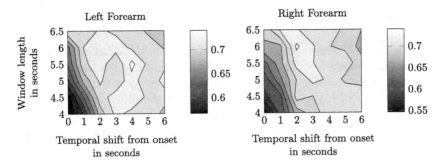

Fig. 7. Video signal window assessment (Baseline temperature (T_0) vs Highest heat temperature (T_3)). Participant dependent 10-fold cross validation (median). The best performance is achieved for both datasets with a temporal shift from the onset between 2 and 4 s and a corresponding window length between 5 and 6.5 s.

in time when the targeted heat temperature is attained until the point in time when the decreasing temperature reaches the baseline temperature again. Thus, for the following analysis we chose a window length of 6.5 s with a temporal shift of 2 s from the onset. An overview of the selected windows for both audio and video modalities can be seen in Fig. 8.

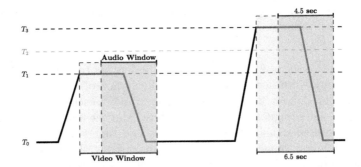

Fig. 8. Audio and video windows. Signal segmentation for feature extraction and classification.

Based on these results additional features were extracted from the specified window. Using the OpenFace toolkit [3], estimates of the head pose consisting of 3 rotation angles and 3 position parameters were extracted. Using the same feature extraction pipeline as the one defined previously for the geometric-based features, window level features were extracted from the signals generated by the head pose parameters.

Moreover, Local Binary Patterns from Three Orthogonal Planes (LBP-TOP) [26] features were extracted. Prior to the extraction of the LBP-TOP features, each 6.5 s window was divided in 3 overlapping segments of 2.5 s, with an overlap of 0.5 s between each consecutive segments. From each segment, the LBP-TOP features were extracted and the final feature vector representing an entire window was obtained by concatenating the LBP-TOP features extracted from each segment. Within each segment, each facial region was divided in a 4 × 4 grid of cells with a 25% overlap from one cell to the next. From each resulting cuboid a uniform LBP-TOP feature vector was extracted. These feature vectors were subsequently concatenated to form the segment level feature vector.

Finally, Pyramid Histogram of Oriented Gradients (PHOG) [4] features were also extracted. From each frame in the window, a 3 levels PHOG feature vector with 20 bins was extracted from the facial region. The feature for the whole window was subsequently generated by performing a max pooling from the frame level feature vectors for the entire window of analysis.

The assessment of the extracted features was also performed through a participant specific 10-fold cross validation classification using the data specific to both the baseline temperature (T_0) and the highest heat temperature (T_3), using a Random Forest classifier with its parameter optimised for each specific feature set. Figure 9 depicts the results of the feature assessment. The LBP-TOP features as well as the facial landmarks features (geometric-based features) outperform both PHOG and head pose features. LBP-TOP features perform best and yield a mean accuracy of 74.14% for the right forearm dataset and 72.94% for the left forearm. Facial landmarks come next and yield a mean accuracy of 73.15% for the right forearm and 72.64% for the left forearm. However the best performance is attained by the early fusion of all the extracted features that yields a mean accuracy of 75.86% on the right forearm dataset and 74.55% on the left forearm dataset.

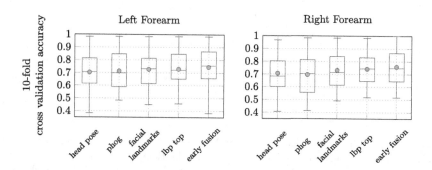

Fig. 9. Video features performance analysis (Baseline temperature (T_0) vs Highest heat temperature (T_3)). Participant dependent 10-fold cross validation. Within each box plot the mean of the classification accuracy across all 40 participants is depicted as a gray dot and the median as a red line. (Color figure online)

Nonetheless, a great variance can be observed across the 40 participants. As pointed out in Sect. 3.2, the level of expressiveness of each participant affects the performance of the system. While responsive participants would display visible facial expressions when submitted to the highest heat temperature, unresponsive participants would not react at all. For such participants the recognition system would perform poorly. An issue may also be the individual heat temperature calibration process, since several participants reported after the experiments not being able to feel any pain at all due to the low stimulation temperatures.

5   Fusion Experiments and Results

Following the assessment of the extracted features from both audio and video channels, several fusion architectures were experimented with in order to investigate the discriminative power of the combined modalities. Since a voice activity detection algorithm was applied on the audio channel to detect voice active segments and silent segments, the audio features are not always available at every time step in comparison to the video channel. Moreover, the length of the voice active segments is not constant and varies greatly from one audio window to the next. Thus, an early fusion of the extracted features from both channels is infeasible. Therefore, the fusion architectures that have been experimented with are late fusion architectures. The fusion is performed using the features of the modalities available at each time step. For each window, if the information from the audio channel is not available, the classification is performed with the features of the video channel uniquely.

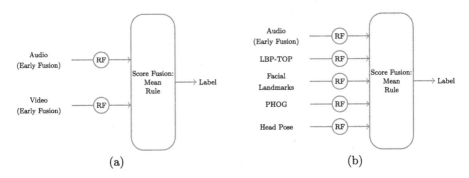

(a)   (b)

Fig. 10. Fusion architectures. (a) a Random Forest classifier is trained on each early fused feature set from both audio and video modalities. The scores of both trained models are combined using the mean rule. (b) in this case a Random Forest classifier is trained on each extracted video feature set and on the early fused audio feature set. The mean rule is subsequently used to fuse the scores of the trained models.

Figure 10 depicts both fusion architectures that have been tested within the scope of the current work. In Fig. 10(a), the extracted features from each modality are early fused and a Random Forest classifier is trained on each of the generated feature sets. A simple mean (average) rule [20] is subsequently applied on the classification scores of both models in order to assign a label to an unseen window. In Fig. 10(b) the features extracted from the video channel are not early fused. Instead, a Random Forest classifier is trained and optimized on each video feature set as well as on the early fused audio feature set, before a simple mean rule is applied on the scores of the individual models to assign a label to an unseen window. The fusion architectures are subsequently tested in a "No Pain" vs "Pain" scenario. We train and test the fusion architectures with the data specific to the baseline temperature (T_0) in combination with the data specific to each heat temperature (T_1, T_2, T_3) successively. The performances of the fusion architectures are compared with the performances yielded by both audio and video modalities when the extracted feature vectors are early fused. Figure 11 depicts the results of the participant dependent scenario.

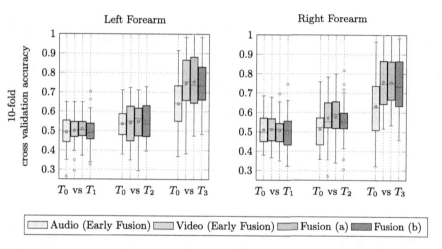

Fig. 11. Fusion architecture assessment ("No Pain" vs "Pain"). Participant dependent 10-fold cross validation. Within each box plot the mean of the classification accuracy across all 40 participants is depicted as a gray dot and the median as a blue line. (Color figure online)

The first observation is the fact that the average accuracy of the classification task increases with the pain intensity in both experimental settings (left and right forearms) and for each tested classification system. However, lower pain intensities which correspond to the temperatures T_1 and T_2 are very difficult to discriminate from the baseline temperature, since the best classification performances in both experimental settings for such pairings $(T_0$ vs T_1, T_0 vs $T_2)$ are barely above chance. These findings can be explained by the fact that high pain intensities cause more observable reactions in both audio and video channels, which results into better classification performances.

Secondly, the video channel outperforms the audio channel in every experimental setting. This can be explained by the fact that the classification task is performed on breathing and moaning recordings. These recordings do not carry

Table 1. Left forearm: participant dependent classification performance.
The mean classification accuracy as well as the standard deviation for 40 participants is depicted for each classification task. The first fusion architecture (see Fig. 10(a)) outperforms the other classification systems. Still, its performance is not significantly better than the one based uniquely on the video channel.

Pairing	T_0 vs T_1	T_0 vs T_2	T_0 vs T_3
Audio (early fusion)	49.29% (±0.079)	53.61% (±0.082)	64.08% (±0.148)
Video (early fusion)	50.14% (±0.074)	54.1% (±0.102)	74.55% (±0.143)
Fusion (a)	**50.91%** (±0.062)	**54.96%** (±0.098)	**75.44%** (±0.141)
Fusion (b)	49.93% (±0.073)	54.6% (±0.096)	74.04% (±0.135)

as much relevant information as the facial region for the classification task and thus perform worse. Moreover, Tables 1 and 2 depict the classification results in form of average accuracy and standard deviation for each pairing and each tested classification system. For the dataset specific to the left forearm, the late fusion architecture consisting of fusing the scores of models trained on early fused features from both audio and video modalities (see Fig. 10(a)) yields the best performance for each pairing, followed by the video channel. For the pairing T_0 vs T_3 a maximum average classification accuracy of 75.44% could be attained. Still, after investigating the significance of the results by using a Wilcoxon sign rank test it was found that the fusion architecture does not outperform the video channel significantly.

Table 2. Right forearm: participant dependent classification performance.
The mean classification accuracy as well as the standard deviation for 40 participants is depicted for each classification task. The classification system based uniquely on the video channel performs best in most cases but still not significantly, in comparison to the first fusion architecture (see Fig. 10(a)).

Pairing	T_0 vs T_1	T_0 vs T_2	T_0 vs T_3
Audio (early fusion)	50.87% (±0.077)	51.39% (±0.091)	63.25% (±0.155)
Video (early fusion)	**51.17%** (±0.078)	57.23% (±0.115)	**75.86%** (±0.134)
Fusion (a)	50.65% (±0.082)	**57.67%** (±0.111)	75.57% (±0.134)
Fusion (b)	49.61% (±0.091)	55.71% (±0.095)	74.47% (±0.141)

Concerning the data specific to the right forearm, the video channel performs the best for the pairings T_0 vs T_1 and T_0 vs T_3, but still not significantly in comparison to the first fusion architecture. The latter performs best for the pairing T_0 vs T_2. A maximum average classification accuracy for the pairing T_0 vs T_3 of 75.86% could be attained.

Finally, the same experiments were performed in a leave one participant out cross validation scenario in order to investigate the power of generalisation of the

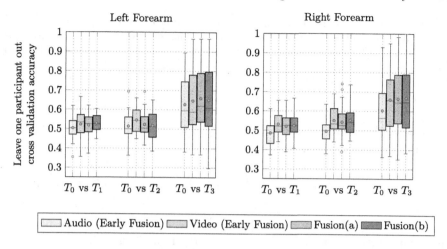

Fig. 12. Fusion architecture assessment ("No Pain" vs "Pain"). Participant independent leave one participant out cross validation. Within each box plot the mean of the classification accuracy across all 40 participants is depicted as a gray dot and the median as a blue line. (Color figure online)

extracted features as well as the generalisation performance of the designed fusion architectures. Figure 12 depicts the performance of each classification architecture for each pairing. Identical to the participant dependent scenario, the higher the considered heat temperature the better the classification performance. Lower temperatures (T_1 and T_2) are even harder to discriminate from the baseline temperature (T_0). Concerning the pairing T_0 vs T_3, both fusion architectures outperform both single modality classification systems but still not significantly (in comparison to the video based classification system).

Tables 3 and 4 depict the classification performances of each system in the form of average accuracy and standard deviation for both left and right forearms datasets. Concerning the pairing T_0 vs T_3, the first fusion architecture (see Fig. 10(a)), performs best with an average accuracy of 65.89% for the left

Table 3. Left forearm: participant independent leave one participant out cross validation classification performance. The mean classification accuracy as well as the standard deviation for 40 participants is depicted for each classification task. Both fusion architectures outperform the single modality systems for the pairing T_0 vs T_3, but still not significantly.

Pairing	T_0 vs T_1	T_0 vs T_2	T_0 vs T_3
Audio (early fusion)	50.37% (±0.054)	51.45% (±0.064)	62.69% (±0.144)
Video (early fusion)	52.33% (±0.066)	**54.59% (±0.062)**	64.59% (±0.158)
Fusion (a)	51.86% (±0.049)	52.31% (±0.061)	**65.89% (±0.172)**
Fusion (b)	**52.90% (±0.043)**	51.78% (±0.065)	65.00% (±0.172)

Table 4. Right forearm: participant independent leave one participant out cross validation classification performance. The mean classification accuracy as well as the standard deviation for 40 participants is depicted for each classification task. Both fusion architectures outperform the single modality systems for the pairing T_0 vs T_3, but still not significantly.

Pairing	T_0 vs T_1	T_0 vs T_2	T_0 vs T_3
Audio (early fusion)	48.59% ($\pm$0.062)	49.55% ($\pm$0.054)	60.35% ($\pm$0.144)
Video (early fusion)	**53.20%** ($\pm$0.057)	55.16% ($\pm$0.069)	65.95% ($\pm$0.165)
Fusion (a)	52.09% ($\pm$0.060)	54.32% ($\pm$0.072)	66.36% ($\pm$0.165)
Fusion (b)	52.50% ($\pm$0.057)	**55.33%** ($\pm$0.073)	**66.76%**($\pm$**0.174**)

forearm dataset, while the second fusion architecture (see Fig. 10(b)) performs best with an average accuracy of 66.76% for the right forearm dataset.

In summary, the classification task becomes very challenging for lower stimulation temperatures, in both participant dependent and independent settings. The video channel outperforms the audio channel significantly in every classification task. Thus, more relevant discriminative information can be extracted from the facial region than in the recorded breathing noises. Still, for the pairing T_0 vs T_3, the audio channel performs significantly better than random classification with performances above 60% in every classification task and setting. Moreover, the considered fusion architectures would outperform both single modality classification architectures given that the performances of the latter are above a certain threshold. Thus, further analyses are to be undertaken to optimize the performance of the fusion architectures by introducing adequate weights and investigating relevant levels of fusion, as suggested in related works [13,23].

6   Conclusion and Future Work

In the present work, a newly recorded dataset in the scope of pain and emotion recognition research has been presented. The first analysis conducted on the dataset consisting of uni-modal and multi-modal pain intensity recognition assessment based on both audio and video channels has been described. The yielded results show that the discrimination from the baseline temperature is easier for higher stimulation temperatures. The video channel outperforms the audio channel in both participant dependent and independent settings. However, the audio channel performs significantly better than average for the pairing T_0 vs T_3 (baseline temperature vs highest heat temperature) and for each setting, thus is relevant for the classification task. The tested fusion architectures improve the results of the uni-modal systems but not significantly in comparison to the video based classification system. However, there is some room for improvement. This has to be done by testing different fusion architectures and by introducing relevant weights for the considered modalities. Moreover, relevant features from the bio-physiology modalities should be extracted and experimented with, in order to improve the classification accuracy in each setting. Furthermore, the classification task should be replaced by a regression task in order to proceed with a continuous evaluation of pain intensities using a combination of the available modalities. Additionally, since the performance of the whole system is affected by the level of expressiveness of each participant, a personalisation scheme [12,14,21] is believed to be able to improve the classification and regression performances. Therefore several personalisation settings should be assessed and experimented with.

Acknowledgments. Viktor Kessler and Friedhelm Schwenker are active within the Transregional Collaborative Research Centre SFB/TRR 62 Companion-Technology for Cognitive Technical Systems, funded by the German Research Foundation (DFG).

References

1. Amirian, M., Kächele, M., Schwenker, F.: Using radial basis function neural networks for continuous and discrete pain estimation from bio-physiological signals. In: Schwenker, F., Abbas, H.M., El Gayar, N., Trentin, E. (eds.) ANNPR 2016. LNCS, vol. 9896, pp. 269–284. Springer, Cham (2016). doi:10.1007/978-3-319-46182-3_23
2. Aung, M.S.H., Kaltwang, S., Romera-Paredes, B., Martinez, B., Singh, A., Cella, M., Valstar, M., Meng, H., Kemp, A., Shafizadeh, M., Elkins, A.C., Kanakam, N., de Rothschild, A., Tyler, N., Watson, P.J., Williams, A.C., Pantic, M., Bianchi-Berthouze, N.: The automatic detection of chronic pain-related expression: requirements, challenges and multimodal dataset. IEEE Trans. Affect. Comput. **7**, 435–451 (2016)
3. Baltrusaitis, T., Robinson, P., Morency, L.P.: OpenFace: an open source facial behavior analysis toolkit. In: 2016 IEEE Winter Conference on Applications of Computer Vision, pp. 1–10 (2016)
4. Bosch, A., Zisserman, A., Munoz, X.: Representing shape with a spatial pyramid kernel. In: Proceedings of the 6th ACM International Conference on Image and Video Retrieval, pp. 401–408 (2007)
5. Breiman, L.: Random forests. Mach. Learn. **45**, 5–32 (2001)
6. Chu, Y., Zhao, X., Yao, J., Zhao, Y., Wu, Z.: Physiological signals based quantitative evaluation method of the pain. In: Proceedings of the 19th IFAC World Congress, pp. 2981–2986 (2014)
7. Eyben, F., Weninger, F., Gross, F., Schuller, B.: Recent developments in openSMILE, the Munich open-source multimedia feature extractor. In: ACM Multimedia (MM), pp. 835–838 (2013)
8. Florea, C., Florea, L., Vertan, C.: Learning pain from emotion: transferred HoT data representation for pain intensity estimation. In: Agapito, L., Bronstein, M.M., Rother, C. (eds.) ECCV 2014. LNCS, vol. 8927, pp. 778–790. Springer, Cham (2015). doi:10.1007/978-3-319-16199-0_54
9. Gruss, S., Treister, R., Werner, P., Traue, H.C., Crawcour, S., Andrade, A., Walter, S.: Pain intensity recognition rates via biopotential feature patterns with support vector machines. PLoS ONE **10**, e0140330 (2015)
10. Hermansky, H., Morgan, N., Bayya, A., Kohn, P.: RASTA-PLP speech analysis technique. In: Proceedings of the 1992 IEEE International Conference on Acoustics, Speech and Signal Processing, pp. 121–124 (1992)
11. Jagan Mohan, B., Badu N., R.: Speech recognition using MFCC and DTW. In: International Conference on Advances in Electrical Engineering (ICAEE), pp. 1–4 (2014)

12. Kächele, M., Amirian, M., Thiam, P., Werner, P., Walter, S., Palm, G., Schwenker, F.: Adaptive confidence learning for the personalization of pain intensity estimation systems. Evol. Syst. **8**, 1–13 (2016)
13. Kächele, M., Schwenker, F.: Cascaded fusion of dynamic, spatial, and textural feature sets for person-independent facial emotion recognition. In: 2014 22nd Internation Conference on Pattern Recognition, pp. 4660–4665 (2014)
14. Kächele, M., Thiam, P., Amirian, M., Schwenker, F., Palm, G.: Methods for person-centered continuous pain intensity assessment from bio-physiological channels. IEEE J. Sel. Top. Signal Process. **10**, 854–864 (2016)
15. Kächele, M., Thiam, P., Amirian, M., Werner, P., Walter, S., Schwenker, F., Palm, G.: Multimodal data fusion for person-independent, continuous estimation of pain intensity. In: Iliadis, L., Jayne, C. (eds.) EANN 2015. CCIS, vol. 517, pp. 275–285. Springer, Cham (2015). doi:10.1007/978-3-319-23983-5_26
16. Kächele, M., Werner, P., Al-Hamadi, A., Palm, G., Walter, S., Schwenker, F.: Bio-visual fusion for person-independent recognition of pain intensity. In: Schwenker, F., Roli, F., Kittler, J. (eds.) MCS 2015. LNCS, vol. 9132, pp. 220–230. Springer, Cham (2015). doi:10.1007/978-3-319-20248-8_19
17. Kaltwang, S., Rudovic, O., Pantic, M.: Continuous pain intensity estimation from facial expressions. In: Bebis, G., et al. (eds.) ISVC 2012. LNCS, vol. 7432, pp. 368–377. Springer, Heidelberg (2012). doi:10.1007/978-3-642-33191-6_36
18. Krothapalli, S.R., Koolagudi, S.G.: Emotion recognition using vocal tract information. In: Krothapalli, S.R., Koolagudi, S.G. (eds.) Emotion Recognition using Speech Features, pp. 67–78. Springer, New York (2013)
19. Krothapalli, S.R., Koolagudi, S.G.: Speech emotion recognition: a review. In: Krothapalli, S.R., Koolagudi, S.G. (eds.) Emotion Recognition using Speech Features, pp. 15–34. Springer, New York (2013)
20. Kuncheva, L.I.: Combining Pattern Classifiers: Methods and Algorithms. Wiley, Hoboken (2004)
21. Meudt, S., Schwenker, F.: On instance selection in audio based emotion recognition. In: Mana, N., Schwenker, F., Trentin, E. (eds.) ANNPR 2012. LNCS, vol. 7477, pp. 186–192. Springer, Heidelberg (2012). doi:10.1007/978-3-642-33212-8_17
22. Olugbade, T.A., Bianchi-Berthouze, N., Marquardt, N., Williams, A.C.: Pain level recognition using kinematics and muscle activity for physical rehabilitation in chronic pain. In: IEEE Proceedings of International Conference on Affective Computing and Intelligent Interaction, pp. 243–249 (2015)
23. Sun, B., Li, L., Zhou, G., Wu, X., He, J., Yu, L., Li, D., Wei, Q.: Combining multimodal features within a fusion network for emotion recognition in the wild. In: Proceedings of the 2015 ACM International Conference on Multimodal Interaction, pp. 497–502 (2015)
24. Werner, P., Al-Hamadi, A., Niese, R., Walter, S., Gruss, S., Traue, H.C.: Towards pain monitoring: facial expression, head pose, a new database, an automatic system and remaining challenges. In: Proceedings of the British Machine Vision Conference, pp. 1–13 (2013)
25. Werner, P., Al-Hamadi, A., Niese, R., Walter, S., Gruss, S., Traue, H.C.: Automatic pain recognition from video and biomedical signals. In: Proceedings of the International Conference on Pattern Recognition (ICPR), pp. 4582–4587 (2014)
26. Zhao, G., Pietikaeinen, M.: Dynamic texture recognition using local binary patterns with an application to facial expressions. IEEE Trans. Pattern Anal. Mach. Intell. **29**, 915–928 (2007)

The SenseEmotion Database: A Multimodal Database for the Development and Systematic Validation of an Automatic Pain- and Emotion-Recognition System

Maria Velana[1], Sascha Gruss[1], Georg Layher[2], Patrick Thiam[2],
Yan Zhang[2], Daniel Schork[3], Viktor Kessler[2], Sascha Meudt[2],
Heiko Neumann[2], Jonghwa Kim[3], Friedhelm Schwenker[2],
Elisabeth André[3], Harald C. Traue[1], and Steffen Walter[1(✉)]

[1] University Clinic of Psychosomatic Medicine and Psychotherapy,
Medical Psychology, Ulm University, 89075 Ulm, Germany
steffen.walter@uni-ulm.de
[2] Institute of Neural Information Processing, Ulm University,
89069 Ulm, Germany
[3] Department of Computer Science, Human-Centered Multimedia,
University of Augsburg, 86159 Augsburg, Germany

Abstract. In our modern industrial society the group of the older (generation 65+) is constantly growing. Many subjects of this group are severely affected by their health and are suffering from disability and pain. The problem with chronic illness and pain is that it lowers the patient's quality of life, and therefore accurate pain assessment is needed to facilitate effective pain management and treatment. In the future, automatic pain monitoring may enable health care professionals to assess and manage pain in a more and more objective way. To this end, the goal of our *SenseEmotion project* is to develop automatic pain- and emotion-recognition systems for successful assessment and effective personalized management of pain, particularly for the generation 65+. In this paper the recently created *SenseEmotion Database* for pain- vs. emotion-recognition is presented. Data of 45 healthy subjects is collected to this database. For each subject approximately 30 min of multimodal sensory data has been recorded. For a comprehensive understanding of pain and affect three rather different modalities of data are included in this study: biopotentials, camera images of the facial region, and, for the first time, audio signals. Heat stimulation is applied to elicit pain, and affective image stimuli accompanied by sound stimuli are used for the elicitation of emotional states.

Keywords: Pain · Heat stimulation · Affective image stimuli · Biopotentials · Video signals · Audio signals · Automatic recognition

© Springer International Publishing AG 2017
F. Schwenker and S. Scherer (Eds.): MPRSS 2016, LNAI 10183, pp. 127–139, 2017.
DOI: 10.1007/978-3-319-59259-6_11

1   Introduction

The proportion of people aged 65 and over increases constantly over the years. Pain is common among older population and can greatly impact older people's quality of life, their physical and psychological functioning and become a barrier to social inclusion. In clinical and care home settings, accurate assessment of pain is essential for successful pain management. A failure to recognize and treat pain can lead to health problems and the unacceptable suffering of elderly. For instance, the lack of sufficient pain management is associated with pathophysiological effects, such as increased blood pressure and heart rate [24]. In general, valid and reliable pain assessment is necessary to facilitate successful pain management without complications [12,20] and enhance quality of life in older adults.

To date, self-reporting is the standard method for assessing pain. However, self-reporting requires the capacity to comprehend the task and to communicate about the experienced pain [32]. This implies that self-report scales are not always a valid and reliable tool for assessing pain in older people, especially in those who have demonstrable cognitive impairment. Moreover, comprehensive pain assessment should be regularly repeated, particularly if the individual is not able to communicate with health care professionals. Previous research on automatic pain recognition is considerable and can be mainly classified into video- and biopotential-based approaches [6,8,11,19,28,29]. However, there is a relative scarcity of studies incorporating both approaches. To the best of authors' knowledge, the study by Walter et al. [27] was the first one that developed a database using biopotential and visual signals. A multimodal pain recognition system incorporating a set of different modalities such as biosignals (e.g., cardiac electrical activity, trapezius muscle activity, skin conductance and respiration), video signals (e.g., facial expressions, skeleton data and head pose) and paralinguistic information is becoming increasingly important for objective and accurate pain assessment. The goal of the SenseEmotion project is the development of an automatic pain- and emotion-recognition system for the successful assessment and effective personalized management of pain in older people. Thus, a multimodal dataset for pain- and emotion-recognition was developed based on physiological, video- and audio-signals. Heat pain was induced experimentally in different levels. In addition, affective image stimuli selected from the International Affective Picture System (IAPS) [16] and the Emotional Picture Set (EmoPicS) [30] were used to evoke positive, negative and neutral emotions. The image stimuli were accompanied by affective sound stimuli so as to intensify affective reactions that would be induced by the image stimuli.

The aim of the present study was to detect patterns of heat pain intensities under the influence of emotional stimuli. The SenseEmotion Database contains the below unique parameters:

- Highly computer-controlled pain stimulation
- Affective induction through emotional stimuli in a two-dimensional space determined by pleasure and arousal ratings

- Physiological measures - i.e., skin conductance level (SCL), electrocardiography (ECG), electromyography (EMG) and respiration (RSP)
- Multiple camera setup
- Depth map video from a Microsoft Kinect V2 with integrated microphone
- Digital wireless headset microphone in combination with a directional microphone

2   Methodology

2.1   Participants

A total of 45 healthy subjects participated in the experiment and received an expense allowance. Participants were recruited through advertisements placed on the campus of the University of Ulm. All participants were fully informed of the study protocol and provided written informed consent for their participation at the beginning of the study. Subjects were excluded for being <18 years old, having neurological problems, psychiatric disorders, chronic pain, headache disorders, cardiovascular diseases, regular use of analgesic medication, or use of analgesic medication directly before the experiment. The study was conducted according to the ethical guidelines set out in the WMA Declaration of Helsinki (ethical committee approval was granted: 196/10-UBB/bal). The study protocol was approved by the ethics committee of the University of Ulm (Helmholtzstraße 20, 89081 Ulm, Germany).

2.2   Design of the Experiment

The Medoc Pathway thermal stimulator was employed to elicit pain [27]. ATS thermode of 30×30 mm [22] was applied to the forearm of the participant (see Fig. 1). During the entire experiment, participants were seated in a chair with their arm resting on the desk in front of them. This system delivers precise painful and non-painful thermal stimuli [22] under highly controlled conditions without causing tissue damage [18]. Thermal stimuli temperatures range from $32\,°C$ to $55\,°C$ [22]. During the entire experiment, stimuli temperature did not exceed $50.5\,°C$ [28].

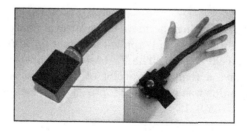

Fig. 1. Thermal pain stimulator that was applied to the participant's forearm. (Reprinted by kind permission of [5], p. 26)

Heat Pain Calibration. At the beginning of the experimental session the researchers determined individual pain threshold (T_1) and pain tolerance threshold (T_3) for each participant. Pain threshold indicates the temperature that the subject's perception alters from heat to pain. Subjects were instructed to press the stop button as soon as heat stimulus became painful: *Please press immediately the stop button, when a feeling of burn, sting, drill, or draw appears in addition to the feeling of heat.* Pain tolerance threshold indicates the temperature that the subject's perception alters from heat to pain and the level that pain becomes intolerable. Hence, subjects were instructed to press the stop button as soon as heat stimulus was barely tolerable: *Please press immediately the stop button, when you cannot accept heat with regard to the feeling of burn, sting, drill, or draw any more.* In order to measure thresholds T_1 and T_3, temperature was gradually raised ($1\,°C/s$) with a starting value of $32\,°C$ (T_0) (see the below section). The researchers performed four measurements for T_1 and T_3 respectively. An average value was calculated for T_1 and T_3 thresholds for each participant. Subsequently, the researchers calculated the mean value of T_1 and T_3 determining one additional individual level T_2 (see Fig. 2). After the calibration phase, Pathway software was calibrated with the three pain levels (T_1, T_2 and T_3) separately for each participant for the main part of the experiment.

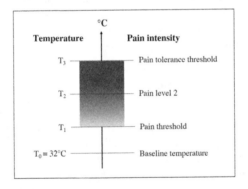

Fig. 2. Induced pain intensity depending on temperature. T_0 represents baseline temperature. T_1, T_2 and T_3 represent the three pain levels that were separately calculated for each participant.

Heat Pain Stimulation. During the main experimental phases each of the three individualized stimuli was randomly applied 30 times for approximately 30 min, resulting in a total of 90 stimuli. The T_0 baseline temperature (no pain) was $32\,°C$. Figure 3 displays a temperature plot of a stimulus and the subsequent pause. Each pain stimulus was performed for 4s. The pauses between the stimuli were randomized between 8–13 s.

Affective Image Stimuli. 180 digital images were selected according to three levels of affective valence (highly pleasant, neutral and highly unpleasant) and two levels of arousal (low, high). During each experimental phase 30 pleasant

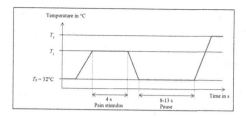

Fig. 3. An example of a 4 s pain stimulus that represents T_1 (pain threshold), and the subsequent pause.

(i.e., erotic and sport categories), 30 unpleasant (i.e., fear and disgust categories) and 30 neutral images were presented under the three pain levels. In total 90 image stimuli were presented with pain stimulation and 30 image stimuli were displayed without pain to every participant in each experimental phase. For each pain level and the baseline temperature all presented images were selected by stratified randomization out of the three categories (pleasant, unpleasant and neutral). The pleasant and unpleasant images were both high in arousal; the neutral images were low in arousal. 108 images were selected from the IAPS [16][1]. These emotional stimuli have normative ratings on affective valence and arousal. Moreover, these ratings have good stability and covary with physiological events [4,15,17]. These images have been utilized extensively in psychophysiological studies and affective computing research [2,21,23,25,27]. Furthermore, 72 images were selected from the EmoPicS [30][2]. The EmoPicS consists of photographic

---

[1] The IAPS identification numbers for the pleasant images were: 1650, 2216, 4311, 4611, 4658, 4659, 4664, 4676, 4677, 4690, 4694, 4695, 4800, 4810, 5460, 5470, 5626, 5629, 7502, 8030, 8080,8178, 8179, 8180, 8185, 8186, 8191, 8193, 8210, 8251, 8300, 8340, 8341, 8370, 8499, 8501. The identification numbers for the unpleasant images were: 1050, 1052, 1113, 1120, 1201, 1525, 1932, 2811, 3150, 3250, 3400, 3500, 5972, 6021, 6022, 6210, 6212, 6260, 6312, 6315, 6415, 6510, 6530, 6550, 6570, 6821, 8480, 8485, 9050, 9250, 9254, 9300, 9600, 9620, 9622, 9902, 9910, 9921. The identification number for the neutral images were: 5471, 5731, 6150, 7002, 7009, 7025, 7030, 7034, 7035, 7036, 7038, 7040, 7041, 7050, 7052, 7053, 7055, 7056, 7057, 7059, 7090, 7100, 7130, 7140, 7150, 7161, 7170, 7185, 7233, 7235, 7493, 7500, 7547, 7705. Mean valence and arousal ratings across image sets were: pleasant images (valance: M = 6.94, arousal: M = 6.40), unpleasant images (valance: M = 2.72, arousal: M = 6.42), and neutral images (valance: M = 5.03, arousal: M = 2.96).

[2] The EmoPicS identification numbers for the pleasant images were: 006, 008, 028, 043, 050, 052, 053, 055, 056, 057, 058, 059, 061, 062, 063, 064, 065, 066, 067, 069, 070, 071, 075, 078. The identification numbers for the unpleasant images were: 207, 210, 211, 213, 214, 216, 219, 222, 229, 231, 232, 235, 238, 244, 250, 251, 252, 254, 321, 325, 326, 329. The identification numbers for the neutral images were: 123, 125, 127, 185, 188, 195, 196, 277, 281, 301, 302, 318, 335, 341, 342, 349, 354, 356, 365, 368, 371, 372, 373, 374, 376, 377. Mean valence and arousal ratings across image sets were: pleasant images (valance: M = 6.94, arousal: M = 5.57), unpleasant images (valance: M = 2.61, arousal: M = 6.26), and neutral images (valance: M = 5.02, arousal: M = 2.84).

affective stimuli with different content and validated normative ratings which are expected to trigger primary motivational states as defined by the theoretical concept of motivational systems [15]. The EmoPicS was developed to serve as a supplement to IAPS. Each picture was presented on a computer screen for 6 s while the individual levels of pain were induced, and SCL, cardiac electrical activity, EMG activity and respiration were measured; moreover, video and audio signals were recorded. The order of the presentation was random.

Affective Sound Stimuli. Each image stimulus was accompanied by a sound stimulus. In the present experiment, the sound stimuli were selected to intensify emotional responses that would be induced by the image stimuli. To this end, sounds were carefully matched to image stimuli in regard to affective valence and arousal. For example, an image of a barking dog was accompanied by an aggressive growling. Each sound stimulus was presented along with the corresponding picture for 6 s over headphones.

2.3   Measured Parameters

Physiological measures and self-report were collected. Biosignal and event data were recorded via Social Signal Interpretation (SSI) [26] (see Fig. 4). SSI provides a flexible open source framework to apply on-the-fly signal processing and pattern recognition to extract higher-level information in real-time. Apart from physiological sensory, a wide range of devices are supported, including audio-visual sensory, motion capture suits, data gloves, pressure-sensitive mats, etc. A core task of SSI is the creation of multimodal databases. To this end, SSI supports the realization of complex multimodal recording setups, possibly distributed over several machines in a network and mechanisms to keep captured data in synchronization without time-stamping. An easy-to-use text-based interface allows users to set up dedicated systems for recording and analyzing multimodal signals without demanding any programming skills.

Biopotentials. g.MOBIlab+ (multi-purpose version) was utilized to acquire EMG and ECG, g.GSRsensor was used to measure SCL. Piezo-electric crystal sensor was utilized to record chest respiration waveforms (for further information: www.gtec.at/Products/Electrodes-and-Sensors/g.Sensors-Specs-Features). The following physiological parameters were measured.

SCL. Skin conductance is a measure of the conductivity of the skin, which especially increases if the skin becomes sweaty [7]. Two electrodes were positioned on the volar pads of the distal phalanges of the index and ring finger to measure SCL. Electrodermal activity is considered to be a sensitive indicator of the *inner tension* of an individual because sweat glands are innervated by the sympathetic branch of the autonomic nervous system (ANS). For instance, a rapid increase in skin conductance response can be reproduced within 1–3 s by the exposure to a stress stimulus - e.g., emotional arousal or intense mental effort.

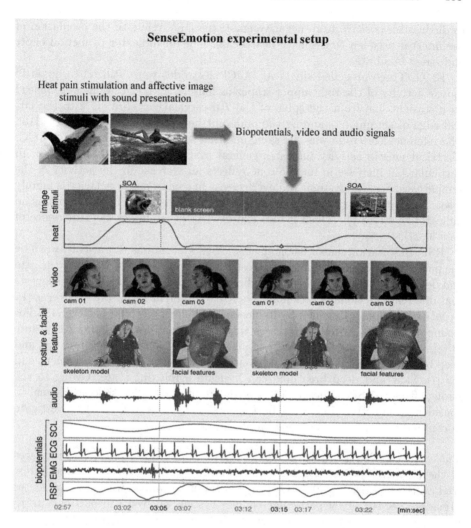

Fig. 4. Overview of the main experimental phase and the parameters that were measured. First row: the right affective stimulus was selected from the IAPS; the identification number of the image was 8186. Second row: the left affective stimulus was selected from the EmoPicS; the identification number of the image was 325. The right affective stimulus was selected from the IAPS; the identification number of the image was 7036.

ECG. Three pregelled single Ag/AgCl electrodes were utilized to measure the average cardiac action potential on the skin. One electrode was placed below the right clavicle (2nd interspace, right midclavicular line). The second electrode was placed on the left lower rib cage (8th interspace, left midclavicular line). The reference electrode was placed on the C7 spinous process. Common features of the ECG signal are heart rate, interbeat interval and heart rate variability (HRV). Heart rate reflects emotional activity [13]. In general, it has been used to distinguish positive from negative emotions. HRV refers to the oscillation of the interval between consecutive heartbeats. It is an indicator of mental effort and stress in adults.

EMG. Three pregelled single Ag/AgCl electrodes were utilized to quantify muscle activity of the right upper trapezius muscle. Two electrodes were placed on a straight line from the spine of the 7th cervical vertebra (C7) to the lateral edge of acromion spanning the midpoint between the two landmarks [10]. The reference electrode was the same one that was used as reference for ECG. Electrical muscle activity indicates general psychophysiological arousal [6]. In particular, an increase in muscle tone reflects an increase in the activity of the sympathetic nervous system, while a decrease in somatomotor activity is mainly associated with parasympathetic arousal. The high level of muscle tension is an indicator of stress [7], which is also expected to occur under the experience of pain [6,27].

RSP. Respiration sensor using an elastic belt system was thoracically worn by the participants over clothing. The most common measures of RSP are the rate and depth of breathing. Evidence for links between emotions and RSP suggests that different emotional states may give rise to different respiratory patterns [1]. Negative emotions can generally induce an irregular breathing pattern [13]. For example, RSP rate usually decreases with relaxation; however, shock exposure or tense situations may cause momentary RSP cessation. Furthermore, fast and deep breathing can be an indicator of emotional arousal such as excitement and joy [7]. Slow and deep breathing can indicate a relaxing state. Owing to the strong effect of RSP on heart rate, RSP is an interesting physiological signal to consider for affective computing both as a signal on its own and to investigate in conjunction with cardiac function [9].

Video Signals. A synchronized camera system was utilized to capture the faces of the participants (see Fig. 4). The camera system consisted of three industrial cameras (iDS UI-3060CP-C-HQ), which were equipped with identical lenses (Tevidon 1.8/16). This particular camera system allowed the participants to move their heads freely ensuring that their faces were fully visible even in case of large out of plane rotations. One camera was placed directly in front of the participant and two at the side. The left and right cameras captured a frontal face in case the participant turned their head 45° to the left or right, respectively. Each camera was connected to a dedicated recording computer via a USB 3.0 cable. Synchronization was realized by externally triggering the three cameras using the ChibiOS real time operation system running on the Arduino Due platform. The implemented setup was capable of constraining the temporal differences of the captured frames within tens of microseconds. Before the recording of participants, the researchers used a checkerboard pattern to calibrate the three cameras. When the recording started, the SSI software [26] first triggered the Arduino board. Then, the board sent triggering signals to all the cameras frequently at a pre-specified frame rate. To optimize the illumination condition, three large LED panels surrounding the participant were used. For the recordings the researchers set the frame resolution to 1600 × 1200, the frame rate to 60 fps for the first 24 recording sessions and to 30 fps for the remaining 21 recordings. Exposure time was set to 15 ms. To convert the raw data to accessible video files, the researchers first performed demosaicing using the built-in function in the OpenCV library and used the codec H.264 to compress the data losslessly. For missing frames due to data transfer failure they reconstructed these frames via temporal interpolation according to the camera time stamps. Additionally, the setup featured a Microsoft Kinect V2. This device was used to record color images at a resolution of 1920 × 1080 pixels at 30fps. The built in depth camera using a time-of-flight sensor was able to provide additional information about the participant. The researchers recorded skeleton data in 3D describing the position of 25 joints, head pose on three axes, 1347 face points in 3D and in 2D for projection on the color image and 17 facial action units at 30 Hz.

Audio Signals. The audio recordings were performed using primarily a digital wireless headset microphone (Line6 XD-V75HS) in combination with a directional microphone (Rode M3). The wireless headset allows unconstrained head movements and records any sound produced by the participants. Typical sounds recorded during these experiments are breathing, moaning and sighing sounds. Meanwhile, the directional microphone records the ambient acoustic sounds. An additional audio stream was recorded using the Microsoft Kinect V2 integrated microphone. This microphone also records the ambient acoustic sounds. All recordings were performed with the sample rate set to 48 kHz. The three audio streams were synchronously recorded with the video and bio-physiological streams using the SSI framework [26].

Visual Analogue Scale. To assess the consistency of subjective criteria utilized to indicate T_1 and T_3 participants rated their pain intensity on the Visual Analogue Scale (VAS) immediately after the end of each experimental phase. The VAS consists of a 100 mm line whose anchors range from *no pain sensation* (score of 0) to *the most intense pain sensation imaginable* (score of 100) [31]. Hence, a higher score points out greater pain intensity. The VAS was administered as a paper-and-pencil measure. Participants were asked to mark on the VAS horizontal straight line the point that they felt best represented their pain intensity.

Self-Assessment Manikin. To verify whether image stimuli elicited the intended emotional states, participants rated their subjective reaction to the induction of affect using the Self-Assessment Manikin (SAM) [3,14] which measures the valence and arousal with viewing each picture. It was emphasized that the researchers were interested in personal feelings and that correct or wrong answers were not possible. The scale includes two sets of five pictographs showing affective valence (unpleasant-pleasant) and arousal (calm-excited). The pleasure scale depicts a smiling figure at one extreme and a frowning figure at the other. The arousal scale represents a sleeping figure at the calm end and an excited and wide-eyed figure at the other. Participants using the mid-point of each scale indicate feeling neither happy nor unhappy, or neither calm nor excited (i.e., neutral). Both scales yielded ratings between 1 and 9 for each dimension, with higher scores being associated with greater subjective pleasure and arousal (lower scores in pleasure dimension are associated with greater displeasure). Ratings were made during picture presentation after the end of each experimental phase. The paper-and-pencil version of the SAM scale was utilized for the present experiment. Participants were instructed to make a mark under the scale of each dimension.

2.4 Procedure

The experiments were conducted at the Emotion Lab[3], Department of Psychosomatic Medicine and Psychotherapy, University of Ulm. First, the researchers summarized each aspect of the content and procedures of the experiment, and obtained informed consent. Each participant's health status was determined by a brief interview to define eligibility. If the participant was eligible, they were seated in a sound and light attenuated room. Then, pain stimulator and physiological sensors were applied. The sites were first slightly abraded with a skin preparation gel and decreased with alcohol for attaching the physiological sensors. Participants were acclimated to the experimental context while they completed several questionnaires. They were informed that no known risk was associated with the procedures of the experiment but they might feel temporary discomfort during skin preparation for the sensors and during heat pain stimulation. The researchers emphasized that the discomfort would be temporary and under participant's control because they would define their pain threshold and pain tolerance threshold. Additionally, participants were told that they were able to leave the experiment at any time by pressing the emergency stop button.

The first 25 participants began with the right arm; the rest of them began with the left arm. For both arms T_1, T_2 and T_3 pain levels were determined and finally both arms of each participant were stimulated with heat pain. The experiment was organized into two phases as shown in Fig. 5. Phase 1 involved calibration phase that lasted 15 min. The main experimental phase involved inducing the individual pain levels while the participant was viewing the image presentation and listening to sound stimuli. The specific procedure was as follows: the main experimental phase began with a preparatory cue which stayed on the screen until the participant adjusted the volume of the headphones for listening to sound stimuli. Sound test was followed by instructions for the participant. Participants were informed that a series of images would be showed on the screen during which they would need to view each image the entire time and listen to the respective sound allowing themselves to experience the potential emotions evoked by the stimuli; they were also instructed that pain stimuli might be induced during the presentation. After explaining the procedure, the researchers left the experimental room and monitored the participants by video camera from a control room. The main experimental phase lasted approximately 30 min. This phase was followed by an after-rating condition. During this condition, each participant rated the intensity of T_1 and T_3 on the VAS. They rated simultaneously

---

[3] http://www.uni-ulm.de/~hhoffman/emotions.

valance and arousal of six affective stimuli that aimed to evoke positive, negative and neutral emotions using the SAM; they also rated three images from the three sets of emotions without pain stimulation. The after-rating condition began with a 6 s presentation of the to-be-rated picture and a 4 s induction of the pain stimulus, directly after which the rating was made. The rating period was 30s, allowing sufficient time for ratings. There was an interval of 20 min between Phase 1 and 2. The same process was followed for Phase 2. After the end of Phase 1 and 2, participants were requested to apply a cold compress to the area of heat pain stimulation for at least 5 min. At the end of the experiment, sensors were removed and participants were debriefed and thanked.

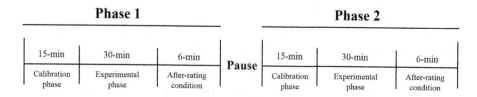

Fig. 5. Experimental procedure

3   Conclusion

The goal of the SenseEmotion project is the development of an automatic pain- and emotion-recognition system for the successful assessment and effective personalized management of pain in elderly. For this purpose, the present study was designed to gather multiple sources of information under heat pain stimulation and viewing image stimuli along with listening to sound stimuli aiming at emotion activation. The SenseEmotion Database consists of biopotentials (i.e., SCL, ECG, trapezius muscle EMG and RSP), video (facial expressions, skeleton data and head pose) and audio (paralinguistic information) signals. The data will be pre-analyzed with a variety of complex filter and decomposition techniques to extract and select meaningful feature patterns that will contribute to the highest recognition rate for pain- and emotion-recognition, pain quantification and differentiation between pain and emotion. Next steps will involve analyzing the data utilizing machine learning algorithms for offline and online analysis for pain- and emotion-recognition in real-time. Further, the researchers plan to advance the present study with the following key aspects:

1. The researchers will test and improve the generalizability with a complex pain model including phasic and tonic pain, heat pain and electrocutaneous stimulation.
2. The research group plans to test the automatic pain- and emotion-recognition system in clinical practice. One idea could be that the automatic recognition system would be tested in a post-operative setting of a care unit for people with dementia syndrome.

To sum up, the SenseEmotion project advances towards its vision of an automatic pain- and emotion-recognition system that will facilitate pain assessment and management in older people in clinical and care home settings.

Acknowledgment. This paper is based on work done within the project *SenseEmotion* funded by the German Federal Ministry of Education and Research (BMBF).

References

1. Boiten, F.A., Frijda, N.H., Wientjes, C.J.E.: Emotions and respiratory patterns: review and critical analysis. Int. J. Psychophysiol. Official J. Int. Organ. Psychophysiol. **17**(2), 103–128 (1994)
2. Bradley, M.M., Codispoti, M., Cuthbert, B.N., Lang, P.J.: Emotion and motivation I: defensive and appetitive reactions in picture processing. Emotion (Washington, D.C.) **1**(3), 276–298 (2001)
3. Bradley, M.M., Lang, P.J.: Measuring emotion: the self-assessment manikin and the semantic differential. J. Behav. Ther. Exp. Psychiatry **25**(1), 49–59 (1994)
4. Greenwald, M.K., Cook, E.W., Lang, P.J.: Affective judgment and psychophysiological response: dimensional covariation in the evaluation of pictorial stimuli. J. Psychophysiol. **3**, 51–64 (1989)
5. Gruss, S.: Schmerzerkennung anhand psychophysiologischer Signale mithilfe maschineller Lerner. Dissertation, Universität Ulm (2015)
6. Gruss, S., Treister, R., Werner, P., Traue, H.C., Crawcour, S., Andrade, A., Walter, S.: Pain intensity recognition rates via biopotential feature patterns with support vector machines. PLoS ONE **10**(10), 1–14 (2015)
7. Haag, A., Goronzy, S., Schaich, P., Williams, J.: Emotion recognition using biosensors: first steps towards an automatic system. In: André, E., Dybkjær, L., Minker, W., Heisterkamp, P. (eds.) ADS 2004. LNCS, vol. 3068, pp. 36–48. Springer, Heidelberg (2004). doi:10.1007/978-3-540-24842-2_4
8. Hammal, Z., Cohn, J.F.: Automatic detection of pain intensity. In: Proceedings of the 14th ACM International Conference on Multimodal Interaction, ICMI 2012, pp. 47–52. ACM, New York (2012)
9. Healey, J.: Physiological sensing of emotion. In: Calvo, R., D'Mello, S., Gratch, J., Kappas, A., (eds.) The Oxford Handbook of Affective Computing, pp. 204–216. Oxford University Press, New York (2015)
10. Jensen, C., Vasseljen, O., Westgaard, R.H.: The influence of electrode position on bipolar surface electromyogram recordings of the upper trapezius muscle. Eur. J. Appl. Physiol. **67**(3), 266–273 (1993)
11. Kächele, M., Amirian, M., Thiam, P., Werner, P., Walter, S., Palm, G., Schwenker, F.: Adaptive confidence learning for the personalization of pain intensity estimation systems. Evolving Syst. **8**(1), 71–83 (2017)
12. Kehlet, H.: Acute pain control and accelerated postoperative surgical recovery. Surg. Clin. North Am. **79**(2), 431–443 (1999)
13. Kim, J., André, E.: Emotion recognition based on physiological changes in music listening. IEEE Trans. Pattern Anal. Mach. Intell. **30**(12), 2067–2083 (2008)
14. Lang, P.J.: Behavioral treatment and bio-behavioral assessment: Computer applications. In: Sidowski, J.B., Johnson, J.H., Williams, T.A. (eds.) Technology in Mental Health Care Delivery Systems, pp. 119–137. Ablex Publishing, Norwood (1980)

15. Lang, P.J.: The emotion probe: studies of motivation and attention. Am. Psychol. **50**(5), 372–85 (1995)
16. Lang, P.J., Bradley, M.M., Cuthbert, B.N.: International affective picture system (IAPS): Affective ratings of pictures and instruction manual. Technical report A-8, University of Florida, Gainesville, FL (2008)
17. Lang, P.J., Greenwald, M.K., Bradley, M.M., Hamm, A.O.: Looking at pictures: affective, facial, visceral, and behavioral reactions. Psychophysiology **30**(3), 261–273 (1993)
18. Lautenbacher, S.: Schmerzmessung. In: Basler, H.D., Franz, C., Kröner-Herwig, B., Rehfisch, H.P. (eds.) Psychologische Schmerztherapie, pp. 271–288. Springer, Berlin (2004)
19. Limbrecht-Ecklundt, K., Werner, P., Traue, H.C., Walter, S.: Mimische Aktivität differenzierter Schmerzintensitäten. Korrelation der Merkmale von Facial Action Coding System und Elektromyografie. Der. Schmerz **30**(3), 248–256 (2016)
20. McQuay, H., Moore, A., Justins, D.: Treating acute pain in hospital. Br. Med. J. **314**(7093), 1531–1535 (1997)
21. Meagher, M.W., Arnau, R.C., Rhudy, J.L.: Pain and emotion: effects of affective picture modulation. Psychosom. Med. **63**(1), 79–90 (2001)
22. Medoc advanced medical systems (2009)
23. Rhudy, J.L., Williams, A.E., McCabe, K.M., Nguyen, M.A., Rambo, P.: Affective modulation of nociception at spinal and supraspinal levels. Psychophysiology **42**(5), 579–587 (2005)
24. Serpell, M.: Handbook of Pain Management. Springer, New York (2008)
25. Tan, J.W., Andrade, A.O., Li, H., Walter, S., Hrabal, D., Rukavina, S., Limbrecht-Ecklundt, K., Hoffman, H., Traue, H.C.: Recognition of intensive valence and arousal affective states via facial electromyographic activity in young and senior adults. PLoS ONE **11**(1), 1–14 (2016)
26. Wagner, J., Lingenfelser, F., Baur, T., Damian, I., Kistler, F., André, E.: The social signal interpretation (SSI) framework: multimodal signal processing and recognition in real-time. In Proceedings of the 21st ACM International Conference on Multimedia - MM 2013, pp. 831–834. ACM Press, New York (2013)
27. Walter, S., Gruss, S., Ehleiter, H., Tan, J., Traue, H.C., Crawcour, S., Werner, P., Al-Hamadi, A., Andrade, A.O., Moreira da Silva, G.: The BioVid heat pain database - data for the advancement and systematic validation of an automated pain recognition system. In: 2013 IEEE International Conference on Cybernetics (CYBCONF), pp. 128–131. IEEE, June 2013
28. Walter, S., Gruss, S., Limbrecht-Ecklundt, K., Traue, H.C., Werner, P., Al-Hamadi, A., Diniz, N., Moreira, G., Andrade, A.O.: Automatic pain quantification using autonomic parameters. Psychol. Neurosci. **7**(3), 363–380 (2014)
29. Werner, P., Al-Hamadi, A., Limbrecht-Ecklundt, K., Walter, S., Gruss, S., Traue, H.C.: Automatic pain assessment with facial activity descriptors. IEEE Trans. Affect. Comput. **99**, 1–14 (2016)
30. Wessa, M., Kanske, P., Neumeister, P., Bode, K., Heissler, J., Schönfelder, S.: EmoPics: Subjektive und psychophysiologische Evaluation neuen Bildmaterials für die klinisch-bio-psychologische Forschung. Zeitschrift für Klinische Psychologie und Psychotherapie **39**(Suppl. 1/11), 77 (2010)
31. Wewers, M.E., Lowe, N.K.: A critical review of visual analogue scales in the measurement of clinical phenomena. Res. Nurs. Health **13**(4), 227–236 (1990)
32. Zwakhalen, S.M.G., Hamers, J.P.H., Abu-Saad, H.H., Berger, M.P.F.: Pain in elderly people with severe dementia: a systematic review of behavioural pain assessment tools. BMC Geriatr. **6**(3), 1–15 (2006)

Photometric Stereo for 3D Face Reconstruction Using Non Linear Illumination Models

Barbara Villarini[1]([✉]), Athanasios Gkelias[2], and Vasilios Argyriou[3]

[1] University of Westminster, London, UK
b.Villarini@westminster.ac.uk
[2] Imperial College, London, UK
a.gkelias@imperial.ac.uk
[3] Kingston University, London, UK
Vasileios.Argyriou@kingston.ac.uk

Abstract. Face recognition in presence of illumination changes, variant pose and different facial expressions is a challenging problem. In this paper, a method for 3D face reconstruction using photometric stereo and without knowing the illumination directions and facial expression is proposed in order to achieve improvement in face recognition. A dimensionality reduction method was introduced to represent the face deformations due to illumination variations and self shadows in a lower space. The obtained mapping function was used to determine the illumination direction of each input image and that direction was used to apply photometric stereo. Experiments with faces were performed in order to evaluate the performance of the proposed scheme. From the experiments it was shown that the proposed approach results very accurate 3D surfaces without knowing the light directions and with a very small differences compared to the case of known directions. As a result the proposed approach is more general and requires less restrictions enabling 3D face recognition methods to operate with less data.

Keywords: Face reconstruction · Face recognition · Photometric stereo · 3D imaging · Non-linear dimensionality reduction · Illumination models

1 Introduction

Automatic face and facial expression recognition has become a very vibrant topic in the last decade due to the fast progress of human computer intelligent interaction (HCII). Face recognition on frontal faces under controlled condition, such as known light condition, is a mature research field and high recognition accuracy can be achieved. However, the performance decreases in presence of illuminations changes, pose variations, facial expressions and a large number of subjects [12]. A large amount of interest has been addressed towards 3D modeling and reconstruction of faces in order to improve face and facial expression recognition. Photometric stereo techniques are used to estimate the illumination condition and to extract 3D geometry information of a face [2,17,18].

F. Schwenker and S. Scherer (Eds.): MPRSS 2016, LNAI 10183, pp. 140–152, 2017.
DOI: 10.1007/978-3-319-59259-6_12

Photometric stereo techniques are further categorised into constrained and unconstrained. Unconstrained photometric stereo means that we do not have any priori knowledge of the light-source direction or the light-source intensity. However, in constrained photometric stereo the light-source direction is not an unknown, and for this reason it is easier to determine the surface normal and the surface reflectance. The main issue related to the unconstrained photometric stereo approaches is that a large number of images is required captured while the light source is moving around the observed object. Therefore, this process is not convenient for real time applications since both the capturing and processing time increases significantly. In this paper in order to solve these problems an unconstrained photometric stereo approach for face reconstruction is proposed requiring at least three images of the surface captured under different unknown illumination directions. The primary contributions of this work are twofold. First, we present a novel methodology for illumination direction estimation for human faces based on low dimensional subspaces. Secondly, we perform an analysis for the accuracy of the obtained lighting conditions in terms of reconstruction accuracy and computational complexity.

This paper is organised as follows. In Sect. 2 previous work on photometric stereo and low dimensional subspaces is reviewed. In Sect. 3, we present an overview of the proposed approach. In Sect. 4 experiments are performed using different datasets and metrics indicating the advantages of the proposed approach. Finally, conclusions on the proposed methodology and the evaluation process are presented.

2   Previous Work

Woodham [30] was the first to introduce photometric stereo. He proposed a method which was simple and efficient, but only dealt with Lambertian surfaces and was sensitive to noise. An unconstrained photometric-stereo method for estimating the surface normal and the surface reflectance of objects without a priori knowledge of the light source directions was proposed in [7]. Also, worth mentioning the work in [9,16,29] that are based on similar approaches. Recently approaches for 3D reconstruction based on photometric stereo with unknown lighting were proposed [1,4,21,22,25–27,31]. Regarding all these methods, the main difference is that they require a significant amount of images captured while the illumination direction is changing, which makes them unsuitable for real time applications and scenarios that involve humans due to their unconstrained self-movement.

To the best of our knowledge, few approaches based on low dimensional subspaces have been used on photometrics. However, a few of them have potential since they are able to consider explicitly the modelling of illumination changes in their methodology. Georghiades et al. [9] a generative model is created by using low dimensional linear subspaces in order to reconstruct new poses and illuminations. Although useful for face recognition, the systems is face specific and therefore unable to extrapolate to new subjects as required in photometrics.

Hallinan [10] propose a low-dimensional model for human faces that can both synthesize a face image when given lighting conditions and can estimate lighting conditions when given a face image. Althugh based on PCA, the method is designed to explain lighting conditions and not to discount them. However this is achieve by limiting all the other sources of variability, including the usage of different people, and will fail otherwise. Lee [20] also propose to represent the illumination cone in a low dimensional space, similarly to [9], by using spherical armonics. Again, the generated low dimensional spaces are person specific.

As a common limitation factor, all these approaches rely on linear methodologies, and therefore, on the assumption of linearity in the face subspace. This assumption is only true under certain conditions, such as face specific applications, as we will explain in Sect. 3.1, which lead to poor performances for realistic scenarios with multiple subjects, different illumination conditions and facial expressions. Instead, we propose a non linear low dimensional space able to model these factors and take advantage of it for 3D face reconstruction and related recognition applications.

3   Proposed Methodology

In this work a novel two step approach for face reconstruction using photometric stereo is proposed. Initially, illumination direction is estimated using low dimensional illumination models. During this step a large number of faces illuminated from all possible directions on a hemisphere is used for training and to create a manifold that represents all the lighting directions in a 3D space. In the second part of the proposed methodology, a new face is captured and at least 3 images are obtained illuminated from different unknown directions. The images are transferred to the new space using the mapping function obtained from the previous step and the corresponding illumination direction is obtained. Finally, photometric stereo is applied and the surface normals are estimated. The proposed algorithm is analysed in the following sections.

3.1   Illumination Subspace

Using a set of training images from different people and lighting directions, a low dimensional subspace can be generated [9]. In this subspace, the distribution of the training samples allows us to estimate the illumination direction while the usage of different subjects allow generalisation to new test subjects.

Dimensionality reduction techniques have been used frequently to model face appearance and pose [11], but illumination variations have been rarely considered or accepted as a factor to be removed [9,10,20]. This is mainly due to the false assumption that the low dimensional subspace comprising the face samples is linear. Belhumeur et al. [5] demonstrated that all images of a given Lambertian surfaces, taken from a fixed point of view, and under varying illuminations can lie in a 3D linear subspace, but they also pointed that shadowing, facial expressions and other factors produce that regions of the face may exhibit deviation from a linear subspace. The required usage of several subjects in the dataset, and therefore different Lambertian surfaces, implies a substantial factor that affects this linearity.

In order to deal with all the possible, we propose a nonlinear dimensionality reduction (DR) technique, as oppose to other approaches which try to linearise the space by choosing an ad-hoc optimal linear and discarding non linear factors. Many different non linear DR techniques could be used, both mapping based (GPLVM [19], GPDM [14]) or spectral based (LE [6], LLE [24], Isomap [13]). However, the big differences among the face surfaces can hide the illumination information, being discarded during the DR process by conventional techniques. Among these methodologies, t-Stochastic Neighbour Embedding (t-SNE) [28] has been proposed to provide a mathematical framework where new constrains can easily be introduced.

3.1.1   Illumination Manifold Using Stochastic Neighbour Embedding

t-Stochastic Neighbour Embedding (t-SNE) [28] is a non-linear dimensionality reduction technique used to embed high-dimensional data into a low-dimensional space (e.g., two or three dimensions for human-intuitive visualization). Given a set of N high-dimensional faces of people under different illumination conditions (i.e. data-points) $x_1, ..., x_N$, t-SNE starts by converting the high-dimensional Euclidean distances between data-points ($\|x_i - x_j\|$)into pairwise similarities given by symmetrized conditional probabilities. In particular, the similarity between data-points x_i and x_j is calculated from (1) as:

$$p_{ij} = \frac{p_{i|j} + p_{j|i}}{2N} \tag{1}$$

where $p_{i|j}$ is the conditional probability that x_i will choose x_j as its neighbour if neigbours were picked in proportion to their probability density under a Gaussian centred at x_i with variance σ_i^2, given by (2):

$$p_{i|j} = \frac{\exp\left(-\|x_i - x_j\|^2 / 2\sigma_i^2\right)}{\sum_{k \neq i} \exp\left(-\|x_k - x_i\|^2 / 2\sigma_i^2\right)} \tag{2}$$

In the low-dimensional space the Student-t distribution (with a single degree of freedom: $f(x) = 1/(\pi(1 + x^2))$) that has much heavier tails than a Gaussian (in order to allow dissimilar objects to be modelled far apart in the map) is used to convert distances into joint probabilities. Therefore, the joint probabilities q_{ij} for the low-dimensional counterparts y_i and y_j of the high-dimensional points x_i and x_j are given by

$$q_{ij} = \frac{(1 + \|y_i - y_j\|^2)^{-1}}{\sum_{k \neq l}(1 + \|y_k - y_l\|^2)^{-1}}. \tag{3}$$

The objective of the embedding is to match these two distributions (i.e., (1) and (2)), as well as possible. This can be achieved by minimizing a cost function which is the Kullback-Leibler divergence between the original (p_{ij}) and the induced (q_{ij}) distributions over neighbours for each object

$$D_{KL}(P\|Q) = \sum_i \sum_j p_{ij} \log \frac{p_{ij}}{q_{ij}}. \tag{4}$$

The minimization of the cost function is performed using a gradient decent method which have the following simple form:

$$\frac{\delta D_{KL}}{\delta y_i} = 4 \sum_j \frac{(p_{ij} - q_{ij})(y_i - y_j)}{(1 + \|y_i - y_j\|^2)} \tag{5}$$

3.1.2 Mapping Functions

Once the low dimensional illumination space is obtained, a mapping mechanism is needed in order to project a new face or a subset of them and estimate the most likely illumination direction. Methods such as t-SNE allow unsupervised generation of embedded spaces, but they do not provide explicitly any mapping mechanism between the low and high dimensional spaces. This issue has been tackled very effectively by Radial Basis Function Networks (RBFN) [8]. Projection functions are produced by training direct ϕ and inverse ϕ' sets of functions between high and low dimensional spaces.

$$\phi : \mathbb{R}^N \to \mathbb{R}^n \text{ and } \phi' : \mathbb{R}^n \to \mathbb{R}^N \tag{6}$$

In our framework, multi-dimensional Gaussian activation functions ϕ_j (Eq. 7) are employed because of its flexibility and superior performance to fit the subspace.

$$\phi_j = e^{(-(X-\mu_j)^T \cdot \Sigma_j^{-1} \cdot (X-\mu_j))} \tag{7}$$

for $j = 1, ..., n_g$, where X is the input feature vector, n_g the number of Gaussian functions to be discovered and μ_j and Σ_j the mean and covariance respectively of each Gaussian function.

3.1.3 Illumination Estimation

In order to estimate the most likely illumination directions for a new face sample, a nearest neighbour classifier is used in the embedded space. The embedded space is composed of 9 different subspaces, each of them comprising all the possible azimut angles from 0 to 360, at 9 different elevation angles from 10 to 90°. Each subspace generated with t-SNE produces a radial manifold, were people are overlaping and distributed according to both their appearance and the illumination angle of the lighting. The overall embedded space can be represented as a hemisphere compose of radial manifolds (see Fig. 1).

By projecting a new sample in this space and pairing with the nearest neighbour, the illumination direction is estimated as the same belonging to this nearest neighbour. This strategy has been proved enough for providing a sufficient estimation, as depicted in Fig. 1 and in the result section.

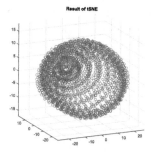
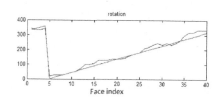

Fig. 1. Left. The obtained 3-dimensional manifold using t-SNE corresponding to the hemispherical embedded space for azimuth and elevation angles. Right. Angular error obtained for this sequence using nearest neighbour classifier for subject 3 projected on illumination subspace composed by subjects [1,2,4–7].

3.2   Photometric Stereo Without Illumination Information

Since a mapping function from the high dimensional space of all possible shadow and highlight deformations that could occur on a human face to a 3D space is obtained, the illumination direction for any face could be estimated using the same mapping mechanism. Regarding the illumination deformations over a face and in general a surface, it is well known that the fraction of light reflected on an object's surface in a certain direction depends upon the optical properties of the surface material. In this paper we use the Lambertian model, thus the fraction of the incident illumination reflected in a particular direction depends only on the surface normals.

All faces share common surface characteristics, which therefore result in similar statistical distributions of normal vectors, and therefore the shadow and highlight deformations share common characteristics too, independent of the human face. This approach can be extended to any class of surface, not only faces, as long as they share similar facet normal distributions.

Based on the Lambertian model that is used, if $\mathbf{n}$ is the normal vector of a surface facet, ρ its albedo with the cosine of the incidence angle θ_i (the angle between the direction of the incident light and the surface normal), $\mathbf{L}$ is the light direction and I the corresponding brightness value recorded for that facet, we have

$$I = \rho \cos(\theta_i) = \rho(\mathbf{L} \cdot \mathbf{n}) \tag{8}$$

Let us now consider a Lambertian surface patch with albedo ρ and normal $\mathbf{n}$, illuminated in turn by several fixed and known illumination sources with directions $\mathbf{L}^1$, $\mathbf{L}^2$, ..., $\mathbf{L}^{\tilde{N}}$. In this case we can express the intensities of the obtained pixels as:

$$I^k = \rho(\mathbf{L}^k \cdot \mathbf{n}), \quad \text{where} \ \ k = 1, 2, ..., \tilde{N}. \tag{9}$$

If we move to a matrix form, Eq. (9) could then be rewritten as

$$\mathbf{I} = \rho[L]\mathbf{n} \tag{10}$$

If there are at least three illumination vectors which are not coplanar, we can calculate ρ and $\mathbf{n}$ using the Least Squares Error technique, which amounts to applying the left inverse of $[L]$:

$$([L]^T[L])^{-1}[L]^T\mathbf{I} = \rho\mathbf{n} \tag{11}$$

Since $\mathbf{n}$ has unit length, we can estimate both the surface normal (as the direction of the obtained vector) and the albedo (as its length). Extra images allow one to recover the surface parameters more robustly.

The problem of reconstructing the surface from the modified normals is considered next and the depth map needs to be obtained. Therefore, the surface is represented as $(x, y, f(x, y))$, and the normal as a function of (x, y) is

$$\mathbf{N}(x, y) = \frac{1}{\sqrt{1 + \frac{\partial f}{\partial x}^2 + \frac{\partial f}{\partial y}^2}} \left(-\frac{\partial f}{\partial x}, -\frac{\partial f}{\partial y}, 1\right)^T \tag{12}$$

To recover the depth map, we need to determine $f(x, y)$ from measured values of the unit normal. There are a number of ways in which a surface may be recovered from a field of surface normals [15, 23]. There are local and global methods based on trigonometry and the minimisation of error functionals, respectively and the most suitable could be selected for this part of the process.

Assume that the measured value of the unit normal at some point (x, y) is $(a(x, y), b(x, y), c(x, y))$. Then

$$\frac{\partial f}{\partial x} = \frac{a(x, y)}{c(x, y)} \qquad \frac{\partial f}{\partial y} = \frac{b(x, y)}{c(x, y)} \tag{13}$$

At this stage we may perform another check on our data set. Because

$$\frac{\partial^2 f}{\partial x \partial y} = \frac{\partial^2 f}{\partial y \partial x} \tag{14}$$

we expect

$$A(x, y) \equiv \frac{\partial\left(\frac{a(x,y)}{c(x,y)}\right)}{\partial y} - \frac{\partial\left(\frac{b(x,y)}{c(x,y)}\right)}{\partial x} \tag{15}$$

to be small (close to zero) at each point.

Assuming that the partial derivatives pass the above sanity test, we can reconstruct the surface up to some constant error in depth. The partial derivatives give the change in surface height with a small step in either the x or the y direction. This means that we can get the surface by summing these changes in height along some path. In particular, we have

$$f(x, y) = \oint_C \left(\frac{\partial f}{\partial x}, \frac{\partial f}{\partial y}\right) \cdot \mathbf{dl} + c \tag{16}$$

where C is a curve starting at some fixed point and ending at (x, y), $\mathbf{dl}$ is the infinitesimal element along the curve and c is a constant of integration, which represents the unknown height of the surface at the starting point. In order to improve further the reconstruction we combine it with a multigrid 2D integration algorithm, which iteratively solves a global minimization problem, and is less sensitive to the propagation of local errors.

4 Experiments and Results

In order to evaluate the accuracy of the proposed methodology a set of experiments was performed using two different datasets and metrics. In more details the first database (see Fig. 2) was captured in our lab and contains 7 faces male and female all of them facing the camera. The second dataset (see Fig. 2) that was used is the Photometric Database presented in [3] captured under a similar setup having more than 300 faces. In both cases, the persons are assumed to be still during the acquisition stage since a high speed camera was used for the acquisition (i.e. 200 frames per second), eliminating the registration problem. In the second stage of the evaluation procedure, photometric stereo is applied on the input images of the databases and the 3D surface of each face is obtained. The faces are aligned using manually placed markers and an affine transformation algorithm. Since the 3D faces are aligned a Lambertian model is used to generate 2D images and shadow maps illuminated under all possible directions (all the possible azimuth angles from 0 to 360, every 10°, and elevation angles from 10 to 90°, every 10°) on a hemisphere as it is shown in Fig. 3. These images at a resolution of 144 × 144 are using as input at the t-SNE as a training set to generate a embedded low dimensional space representation of the illumination variations and shadow deformations on a human face in a low dimensional space. Consecutively, using the one leave out process, where each subject is removed from the training set dataset and used for testing, the faces are reconstructed using the estimated illumination directions for the input images.

4.1 Illumination Estimation

By using the previously mentioned leave-one-out schema, the performance of the automatic illumination estimation can be evaluated as well as the capability of the embedded space to generalise to new subjects out of the training dataset. Results of the average angular error for each of the datasets are reported in Table 1. Regarding the average angular error it was expected to be in that range since two consecutive illumination sources are in range of 15°, which explains the results indicating that the estimated direction is always in the second-order neighborhood. A particular case for subject 3 is depicted in Fig. 1.

Table 1. Overall performance of the angular estimation provided by the nearest neighbour classifier on the illumination embedded space

	Dataset 1	Dataset 2 [3]
Angular error [Degrees]	32.7536	33.5571

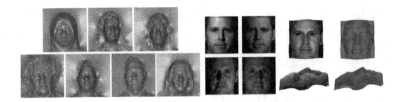

Fig. 2. (Left) An example of faces part of the first dataset. (Right) An example of faces part of the second dataset [3].

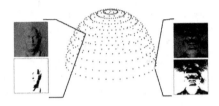

Fig. 3. Examples of possible illumination directions. Each dot on the hemisphere corresponds to a possible light l. It may be identified by its azimuth angle φ_l and its zenith angle θ_l.

It can be concluded that the illumination embedded space is able to provide a reasonably accurate estimation of the lighting direction, which can be now feed into the reconstruction module.

4.2 Reconstruction Performance

In order to evaluate the performance the average difference of the real and the estimated heightmaps was used and furthermore the Hausdorff distance was used to compare the reconstructions with the original profiles.

In more details for the first dataset the results are summarized in Table 2 and examples of the reconstructed surfaces and the height maps are shown in Fig. 4. Also, the error difference between the reconstructed faces knowing the light directions and the one without is shown in Fig. 5. The same experiments were also performed for the second dataset and the obtained results are shown in Table 3 and examples of the obtained surfaces are shown in Figs. 6 and 7. In

Table 2. The accuracy of the proposed photometric stereo method for faces with unknown illumination directions in terms of height map percentage error over the ground truth (case of known directions) for the first dataset.

Average height error	Scenario A	Scenario B
Proposed method	1.4974%	1.1161%
Schindler [25]	2.1775%	1.5553%

Fig. 4. Reconstructed surfaces using the proposed method with unknown illumination on the right and the ground truth on the left under different view points for the first dataset.

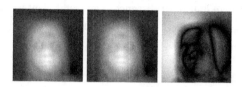

Fig. 5. The error difference between the reconstructed faces knowing the light directions and the one without for the first dataset.

Table 3. The accuracy of the proposed photometric stereo method for faces with unknown illumination directions in terms of height map percentage error over the ground truth (case of known directions) for the second dataset [3].

Average height error	Scenario A	Scenario B
Proposed method	3.0963%	2.7450%
Schindler [25]	3.7831%	2.9806%

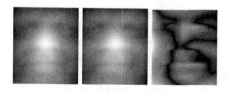

Fig. 6. Reconstructed surfaces using the proposed method with unknown illumination on the right and the ground truth on the left under different view points for the second dataset.

Fig. 7. The error difference between the reconstructed faces knowing the light directions and the one without for the second dataset.

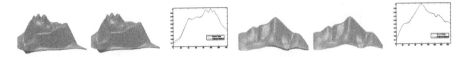

Fig. 8. Examples of profile views used to calculate the Hausdorff distance.

Table 4. The average Hausdorff distance was used to compare the reconstructed with the original profiles.

Average hausdorff distance	Dataset 1	Dataset 2
Proposed method	0.0097	0.0055
Schindler [25]	0.0124	0.0083

more details, for each dataset two tests were performed and in each test scenario three or four different images were used as input to the reconstruction system. The obtained height map in each scenario was compared with the equivalent height map obtained by the same input images but knowing the illumination directions.

The proposed algorithm was further evaluated using the Hausdorff distance comparing the surfaces obtained with and without any illumination information. In Fig. 8 results of the reconstructed faces obtained from the two cases are shown. Observing the results it can be inferred that the proposed methodology results very accurate estimates without knowing the light directions. In particularly, the side view was used to evaluate the performance of the proposed approach. The background was extracted manually and the Hausdorff distance was used to compare the reconstructions with the original profiles. Table 4 shows the average results for all the faces.

5   Conclusions

In this paper, a method for 3D face reconstruction using photometric stereo with unknowing lights was proposed. A dimensionality reduction method was introduced to represent the face deformations due to the illumination variations and the self shadows in a lower space. The obtained mapping function was used to determine the illumination direction of each input image and that direction was used to apply photometric stereo. Experiments with faces were performed in order to evaluate the performance of the proposed scheme in a comparative study. From the experiments it was shown that the proposed approach results very accurate 3D surfaces without knowing the light directions and with a very small differences compared to the case of known directions. As a result the proposed approach is more general and requires less restrictions and information for the acquisition environment, allowing further applications on 3D face recognition and tracking.

References

1. Alldrin, N.G., Mallick, S.P., Kriegman, D.J.: Resolving the generalized bas-relief ambiguity by entropy minimization. In 2007 IEEE Conference on Computer Vision and Pattern Recognition, pp. 1–7, June 2007
2. Argyriou, V., Zafeiriou, S., Villarini, B., Petrou, M.: A sparse representation method for determining the optimal illumination directions in photometric stereo. Signal Process. **93**(11), 3027–3038 (2013)
3. Atkinson, G.A., Hansen, M.F., Smith, W.A.P., Argyriou, V., Petrou, M., Smith, M.L., Smith, L.N.: Face recognition and verification using photometric stereo: the photoface database and a comprehensive evaluation. In: IEEE Transactions on Information Forensics and Security (2013)
4. Basri, R., Jacobs, D., Kemelmacher, I.: Photometric stereo with general, unknown lighting. IJCV **72**(3), 239–257 (2007)
5. Belhumeur, P.N., Hespanha, J.P., Kriegman, D.J.: Eigenfaces vs fisherfaces: recognition using class-specific linear projection. IEEE Trans. Pattern Anal. Mach. Intell. **19**(7), 711–720 (1997)
6. Belkin, M., Nivogi, P.: Laplacian eigenmaps and spectral techniques for embedding and clustering. In: NISP 14 (2001)
7. Chandraker, M.K., Agarwal, S., Kriegman, D.J.: Shadowcuts: photometric stereo with shadows. In: CVPR, June 2007
8. Elgammal, A., Lee, C.: Body pose tracking from uncalibrated camera using supervised manifold learning. In: NIPS EHuM Workshop (2006)
9. Georghiades, A.: Incorporating the torrance and sparrow model of reflectance in uncalibrated photometric stereo. In: 9th ICCV, vol. 2 (2003)
10. Hallinan, P.: A low-dimensional representation of human faces for arbitrary lighting conditions. In: Proceedings of the IEEE Conference on Computer Vision and Pattern Recognition (1994)
11. He, X., Yan, S., Hu, Y., Niyogi, P., Zhang, H.: Face recognition using laplacianfaces. IEEE Trans. PAMI **27**, 328–340 (2005)
12. Hong, J., Song, K.: Facial expression recognition under illumination variation. In: IEEE Workshop on Advanced Robotics and Its Social Impacts, pp. 1–6 (2007)
13. Silva, V., Tenenbaum, J., Langford, J.: A global geometric framework for nonlinear dimensionality reduction. Science **290**(5500), 2319–2323 (2000)
14. Fleet, D., Wang, J., Hertzmann, A.: Gaussian process dynamical models. In: NISP, vol. 18 (2006)
15. Kakadiaris, I.A., Passalis, G., Toderici, G., Murtuza, M.N., Lu, Y., Karampatziakis, N., Theoharis, T.: Three-dimensional face recognition in the presence of facial expressions: an annotated deformable model approach. IEEE Trans. Pattern Anal. Mach. Intell. **29**(4), 640–649 (2007)
16. Kautkar, S.N., Atkinson, G.A., Smith, M.L.: Face recognition in 2d and 2.5d using ridgelets and photometric stereo. Pattern Recogn. **45**, 3317–3327 (2012)
17. Kemelmacher-Shlizerman, I.: Internet based morphable model. In: 2013 IEEE International Conference on Computer Vision, pp. 3256–3263, December 2013
18. Kemelmacher-Shlizerman, I., Seitz, S.M.: Face reconstruction in the wild. In: 2011 International Conference on Computer Vision, pp. 1746–1753, November 2011
19. Lawrence, N.: Gaussian process latent variable models for visualisation of high dimensional data. In: NISP, vol. 16 (2004)
20. Lee, K.C.: Acquiring linear subspaces for face recognition under variable lighting. IEEE PAMI (2005)

21. Lu, F., Matsushita, Y., Sato, I., Okabe, T., Sato, Y.: Uncalibrated photometric stereo for unknown isotropic reflectances. In: 2013 IEEE Conference on Computer Vision and Pattern Recognition (CVPR), pp. 1490–1497, June 2013
22. Papadhimitri, T., Favaro, P.: A closed-form, consistent and robust solution to uncalibrated photometric stereo via local diffuse reflectance maxima. Int. J. Comput. Vis. **107**(2), 139–154 (2014)
23. Robles-Kelly, A., Hancock, E.R.: A graph-spectral approach to shapefrom-shading. IEEE Trans. Image Process. **13**(7), 912–926 (2004)
24. Roweis, S.T., Saul, L.K.: Nonlinear dimensionality reduction by locally linear embedding. Science **290**, 2323–2326 (2000)
25. Schindler, G.: Photometric stereo via computer screen lighting for real-time surface reconstruction. In: Proceedings of the 3DPVT 2008 (2008)
26. Shi, B., Matsushita, Y., Wei, Y., Xu, C., Tan, P.: Self-calibrating photometric stereo. In: 2010 IEEE Conference on Computer Vision and Pattern Recognition (CVPR), pp. 1118–1125, June 2010
27. Shi, B., Wu, Z., Mo, Z., Duan, D., Yeung, S., Tan, P.: A benchmark dataset and evaluation for non-lambertian and uncalibrated photometric stereo. In: IEEE Conference on CVPR, pp. 3707–3716 (2016)
28. van Der-Maaten, G., Hinton, L.: Visualizing data using t-sne. J. Mach. Learn. Res. **9**, 2579–2605 (2008)
29. Argyriou, V., Petrou, M.: Recursive photometric stereo when multiple shadows and highlights are present. In: Proceedings of CVPR (2008)
30. Woodham, R.: Photometric method for determining surface orientation from multiple images. Opt. Eng. **19**(1), 139–144 (1980)
31. Wu, Z., Tan, P.: Calibrating photometric stereo by holistic reflectance symmetry analysis. In: 2013 IEEE Conference on Computer Vision and Pattern Recognition (CVPR), pp. 1498–1505, June 2013

Recursively Measured Action Units

Xiang Xiang[1]([✉]) and Trac D. Tran[2]

[1] Department of Computer Science, Johns Hopkins University,
3400 N. Charles Street, Baltimore, MD 21218, USA
xxiang@cs.jhu.edu
[2] Department of Electrical and Computer Engineering,
Johns Hopkins University, 3400 N. Charles Street, Baltimore, MD 21218, USA

Abstract. Video is a recursively measured signal where frames are highly correlated with structured sparsity and low-rankness. A simple example is facial expression - multiple measurements of a face. Several salient facial action units (AU) are often enough for a correct expression recognition. We hope that AUs are not stored when the face remains neutral until they become salient when expression occurs, as well as that the recognizer is still able to restore historic salient AUs. A temporal memory mechanism is appealing for a real-time system to reduce rich redundancy in information coding. We formulate expression recognition as a video Sparse Representation based Classification (SRC) with Long Short-Term Memory (LSTM) mechanism, which is applicable for human actions yet requiring a careful design of sparse representation due to possible changing scenes. Preliminary experiments are conducted on the MPI Face Video Database (MPI-VDB). We compare the proposed sparse coding with temporal modeling using LSTM against the baseline of sparse coding with simultaneous recursive matching pursuit (SRMP).

1 Introduction

As shown by Fig. 1, the primary problem for a computer to read faces in the wild is the head pose variation [1]. For non-well-aligned faces, it is infeasible to extract a dominant neutral face assumed in the [2]. The solution is to either align faces or design a model robust to pose variation [1]. The way of first aligning the faces and then applying sparse representation counts heavily on the explicit face alignment. In order to relax the constraint of well-aligned faces, we choose to get rid of the low-rank term from the model proposed in [2]. Namely, we hope to represent an expressive face in a certain pose over a dictionary of expressive faces under various poses. As a similar pose or a similar identity can both confuse a similar expression, we rule out the identity by construction of the data. Our observation is that a weighted combination of cropped faces under various poses can also give a probabilistic meaningful face.

As shown in Fig. 2, it is easily observed that several salient facial action units (AU) are often enough for a correct recognition. We hope that AUs are not stored when the face remains neutral until they become salient when expression occurs, as well as that the recognizer is still able to restore historic salient AUs. There exits rich redundancy if we simply code all AUs over time.

© Springer International Publishing AG 2017
F. Schwenker and S. Scherer (Eds.): MPRSS 2016, LNAI 10183, pp. 153–159, 2017.
DOI: 10.1007/978-3-319-59259-6_13

Fig. 1. Faces in the wild are under various poses. Images from the Web.

| Eyebrow Raise | Eyebrow Lower | Smile | Smirk | Nose Wrinkle | Upper Lip Raiser |
| SURPRISE | CONFUSION | AMUSEMENT | SKEPTICISM | DISLIKE/DISGUST | DISLIKE/DISGUST |

Fig. 2. Salient facial action units (AU) are often enough for expression recognition.

2   A Bayes Probabilistic Model

A face $\mathbf{Y}$ is visually affected by confounding factors the primary of which are the identity I, the head pose P and the facial expression E. Thus, we can represent $\mathbf{Y}$ as a function of random variables I, P, E: $\mathbf{Y} = f(I, P, E)$. Then, we denote a specific face as $f(i, p, e) := f(I = i, P = p, E = e)$ given specific i, p, e. If we fix I, the face of a specific person is $f(P, E) := f(I = i, P, E)$. Then, the probability of

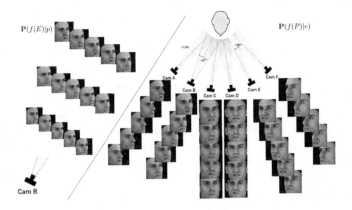

Fig. 3. Illustration of the pose-specific expression model per quantized pose and the expression-specific pose model per quantized expression. Images from the MPI-VDB.

having a specific expression e is $\mathbf{P}(e) := \mathbf{P}(E = e) = \frac{\mathbf{P}(f(P=p,E)|P=p)\cdot\mathbf{P}(P=p)}{\mathbf{P}(f(P)|E=e)} = \frac{\mathbf{P}(f(E)|p)\cdot\mathbf{P}(p)}{\mathbf{P}(f(P)|e)}$ according to the Bayes' Theorem. The likelihood $\mathbf{P}(f(E)|p)$ is modelled using pose-specific expression-varying face videos as shown in Fig. 3-left. The likelihood $\mathbf{P}(f(P)|e)$ is modelled using expression-specific pose-varying face videos as shown in Fig. 3-right. In the following, we take the pose-specific expression case as an example to elaborate the model. The coding for expression-specific poses follows the same algorithm.

3   Coding Pose-Specific Expression with Temporal LSTM

Now, we model an implicit latent representation $\mathbf{X} \in \mathbb{R}^{n\times\tau}$ of an input test face $\mathbf{Y} \in \mathbb{R}^{d\times\tau}$ as a sparse linear combination of prepared fixed training emotions $\mathbf{D} \in \mathbb{R}^{d\times n}$: $\mathbf{Y} = \mathbf{DX}$, where the dictionary matrix $\mathbf{D}$ is an arrangement of all sub-matrices $\mathbf{D}_{[j]}$, $j = 1, ..., \lfloor\frac{n}{\tau}\rfloor$. Notably, n is assumed to be much larger than d and rank($\mathbf{D}$)= d. Namely, our task is to sequentially find a small subset (*i.e.*, basis set) of columns from $\mathbf{D}$ for the Multiple Measurement Vectors (MMV) $\mathbf{X}$.

3.1   Recursive Matching Pursuit Using RNN

As we build up the basis set by adding a single column vector at a time, we denote the residual matrix after the p-th iteration by $\mathbf{R}^{(p)} \in \mathbb{R}^{n\times\tau}$ with $\mathbf{R}^{(0)} = \mathbf{Y}$. The i-th column of $\mathbf{Y}^{(p)}$ is denoted by $\mathbf{y}_i^{(p)}$. The indices of the p vectors selected are stored in the index set denoted by $\mathbb{I}^{(p)}$; where $\mathbb{I}^{(p)} = \{k_1, k_2, ..., k_p\}$ and $\mathbb{I}^{(0)} = \emptyset$. The selected columns vectors are stored in a matrix $\mathbf{S}^{(p)} = [\mathbf{d}_{k_1}, \mathbf{d}_{k_2}, ..., \mathbf{d}_{k_p}]$ and $\mathbf{S}^{(0)} = \emptyset$. The orthogonal projection matrix onto the column space of $\mathbf{S}^{(p)}$ is denoted by $\mathbf{P}_{\mathbf{S}^{(p)}}$ and its orthogonal complement $\mathbf{P}_{\mathbf{S}^{(p)}}^{\perp} = (\mathbf{I} - \mathbf{P}_{\mathbf{S}^{(p)}})$ and $\mathbf{P}_{\mathbf{S}^{(0)}} = \mathbf{0}$, $\mathbf{P}_{\mathbf{S}^{(0)}}^{\perp} = \mathbf{I}$ where $\mathbf{I}$ is an identity matrix and $\mathbf{0}$ is a zero matrix.

The basic idea of the Recursive Matching Pursuit (RMP) algorithm is the pursuit of the matching p-th basis vector conceptually involves solving $(n-p+1)$ order recursive least squares problems and selecting the vector that reduces the residual the most.

We initialize $\mathbf{d}_k^{(0)} = \mathbf{d}_k (\forall k = 1, ..., n)$ and choose a column of $\mathbf{D}$ indexed by

$$k_p = \underset{k}{\arg\max} \left(\sum_{i=1}^{\tau} \left| \left(\mathbf{d}_k^{(p-1)}\right)^T \mathbf{r}_i^{(p)} \right|^2 \Big/ \|\mathbf{d}_k^{(p-1)}\|^2 \right). \tag{1}$$

As $\mathbf{S}^{(p)}$ is augmented, we then update $\mathbf{P}_{\mathbf{S}^{(p)}}$ by

$$\mathbf{P}_{\mathbf{S}^{(p)},k_p} = \mathbf{P}_{\mathbf{S}^{(p)}} + \mathbf{q}^{(p)}\left(\mathbf{q}^{(p)}\right)^T \tag{2}$$

where

$$\mathbf{q}^{(p)} = \frac{\mathbf{d}_{k_p}^{(p-1)}}{\|\mathbf{d}_{k_p}^{(p-1)}\|}. \tag{3}$$

Now we project columns of $\mathbf{D}$ and $\mathbf{R}$ to the column space of $\mathbf{S}^{(p)}$, and get

$$\mathbf{d}_i^{(p)} = \mathbf{P}_{\mathbf{S}^{(p)}}^{\perp}\mathbf{d}_i^{(p-1)} = \mathbf{d}_i^{(p-1)} - \left(\left(\mathbf{q}^{(p)}\right)^T \mathbf{d}_i^{(p-1)}\right)\mathbf{q}^{(p)}, \tag{4}$$

$$\mathbf{r}_i^{(p)} = \mathbf{P}_{\mathbf{S}^{(p)}}^{\perp}\mathbf{r}_i^{(p-1)} = \mathbf{r}_i^{(p-1)} - \left(\left(\mathbf{q}^{(p)}\right)^T \mathbf{r}_i^{(p-1)}\right)\mathbf{q}^{(p)}, \tag{5}$$

respectively, $\forall i = 1, 2, ..., \tau$. Note that no orthogonal projection is employed in the updates. Selection of a column of $\mathbf{D}$ corresponds to selecting a nonzero row of $\mathbf{X}$. The nonzero rows of $\mathbf{X}$ form $\mathbf{X}^{\mathbb{I}}$ which is given by $(\mathbf{S}^{(p)})^{\dagger}\mathbf{Y}$ where $\dagger$ denotes pseudo-inverse.

Algorithm 1. RMP using Recurrent Neural Network.

---

1 function $\mathbf{X} = \text{RMP-RNN}(\mathbf{Y}, \mathbf{D}, resMin)$;

   Input   : measurement matrix $\mathbf{Y} \in \mathbb{R}^{d \times \tau}$, dictionary matrix $\mathbf{D} \in \mathbb{R}^{d \times n}$,
               minimum Frobenius norm $resMin$, trained RNN model.

   Output: sparse-codes matrix $\mathbf{X} \in \mathbb{R}^{n \times \tau}$

2 Initialization $\mathbf{D}^{(0)} = \mathbf{D}$, $\mathbf{X}^{(0)} = \mathbf{0}$, $\mathbf{R}^{(0)} = \mathbf{Y}$;

3 **while** $i \leq \tau$ *and* $\|R\|_F \leq resMin$ **do**

4     $p \leftarrow p + 1$

5     $\mathbf{r}_i^{(p)} \leftarrow \dfrac{\mathbf{r}_{i-1}^{(p)}}{\max\left(|\mathbf{r}_{i-1}^{(p)}|\right)}$

6     $\mathbf{h}_i \leftarrow \text{RNN}(\mathbf{r}_i^{(p)}, \mathbf{h}_{i-1}, \mathbf{c}_{i-1})$

7     $\mathbf{c}_i \leftarrow \text{softmax}(\mathbf{U}\mathbf{h}_j)$

8     $k_p \leftarrow \text{Support}(\max(\mathbf{c}))$

9     $\mathbb{I}^{(p)} \leftarrow \mathbb{I}^{(p-1)} \cup k_p$

10    $\mathbf{S}^{(p)} \leftarrow [\mathbf{S}^{(p-1)}, \mathbf{d}_{k_p}]$

11    $\mathbf{P}_{\mathbf{S}^{(p)}} \leftarrow \mathbf{P}_{\mathbf{S}^{(p)}, k_p}$ by Eq. (2)

12    $\mathbf{d}_i^{(p)} \leftarrow$ Eq. (4)

13    $\mathbf{x}_i^{(p)\mathbb{I}} \leftarrow (\mathbf{S}^{(p)})^{\dagger}\mathbf{y}_i$

14    $\mathbf{r}_i^{(p)} \leftarrow$ Eq. (5)

15 **end**

3.2   RNN Using Long Short-Term Memory (LSTM)

$$\mathbf{h}_i = \mathbf{o}_i * \tanh(\mathbf{f}_i * \mathbf{c}_{i-1} + \mathbf{e}_i * \mathbf{c}_i) \tag{6}$$

where the information memorizing cell is given by

$$\mathbf{c}_i = \tanh(\mathbf{W}_c[\mathbf{h}_{i-1}, \mathbf{r}_i] + \mathbf{b}_c), \tag{7}$$

the vector of information forgetting gates is given by

$$\mathbf{f}_i = \sigma(\mathbf{W}_f[\mathbf{h}_{i-1}, \mathbf{r}_i] + \mathbf{b}_f), \tag{8}$$

the vector of information entering gates is given by

$$\mathbf{f}_e = \sigma(\mathbf{W}_e[\mathbf{h}_{i-1}, \mathbf{r}_i] + \mathbf{b}_e), \tag{9}$$

and the vector of information outputing gates is given by

$$\mathbf{f}_o = \sigma(\mathbf{W}_o[\mathbf{h}_{i-1}, \mathbf{r}_i] + \mathbf{b}_o). \tag{10}$$

4   Related Works

Sparse coding has its root in neuroscience and has been well exploited in harmonic analysis, signal processing and compressive sensing. Matching Pursuit dates back to 1993 in [3] and Recursive Matching Pursuit for sparsely coding Multi-Measurement Vectors can be traced back to 1998 in [4].

Algorithm 2. Baseline: simultaneous recursive matching pursuit (SRMP).

1   function $\mathbf{X} = \text{RMP}(\mathbf{Y}, \mathbf{D}, resMin)$;

   Input : measurement matrix $\mathbf{Y} \in \mathbb{R}^{d \times \tau}$ and minimum Frobenius norm
      $resMin$ as stopping criterion.

   Output: Approximation matrix $\mathbf{A} \in \mathbb{R}^{d \times \tau}$ and a set Λ_p containing p indices
      where p is the number of iterations.

2   Initialization $\mathbf{D}^{(0)} = \mathbf{D}$, $\mathbf{X}^{(0)} = \mathbf{0}$, $\mathbf{R}^{(0)} = \mathbf{Y}$;

3   **while** $i \leq \tau$ and $\|R\|_F \leq resMin$ **do**

4 | $\quad p \leftarrow p + 1$

5 | $\quad k_p \leftarrow$ Eq. (1)

6 | $\quad \mathbb{I}^{(p)} \leftarrow \mathbb{I}^{(p-1)} \cup k_p$

7 | $\quad \mathbf{S}^{(p)} \leftarrow [\mathbf{S}^{(p-1)}, \mathbf{d}_{k_p}]$

8 | $\quad \mathbf{x}_i^{(p)\mathbb{I}} \leftarrow (\mathbf{S}^{(p)})^{\dagger} \mathbf{y}_i$

9 | $\quad \mathbf{d}_i^{(p)} \leftarrow$ Eq. (4)

10 | $\quad \mathbf{r}_i^{(p)} \leftarrow$ Eq. (5)

11 **end**

Recurrent Neural Network (RNN) is deep in time and thus can be treated as a deep neural network with the issue of vanishing gradients. Long Short-Term Memory network [5] dating back to 1997 is a type of RNN that gets around of vanishing gradient problem. Note that there also exists a type of network called Recursive Neural Network [6] which is generalized RNN with a deep structure of a skewed tree. There are connections among all those models and hidden Markov models. Both a feed-forward multilayer neural network and a hidden Markov model can be seen as a directed acyclic graph with hidden nodes. Both a recurrent neural network and a hidden Markov model map a sequence of inputs to a sequence of outputs via a sequence of hidden states. Long Short-Term Memory learns a function of inputs and hidden states using perceptron-like network with gating weights further learned using perceptrons.

5   Experiments

Experiments are conducted on the MPI-VDB consisting of long expression videos simultaneously capture at 6 views. In each video, the same expression repeats several times. Please see http://vdb.kyb.tuebingen.mpg.de for the raw data and https://github.com/eglxiang/FacialAU for the cropped data. Images are cropped using the Viola-Jones face detector. We randomly choose half videos for forming the dictionary and the other half for testing. Figure 4 shows the detected salient AU for different expressions. Figure 5 presents the confusion matrix of sparse coding with LSTM (left) vs. with SRMP (right).

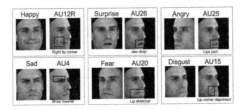

Fig. 4. Detection of salient AU occurrence. Left: the peak frame of an expression sequence at a random view. Right: the peak frame containing a detected salient AU which correctly helps recognizing the expression (shown all at the right profile view).

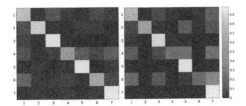

Fig. 5. Confusion matrix of the proposed model (left, with LSTM) and the baseline model (right, with SRMP) on MPI-VDB over 20 runs. There are 7 emotion categories including the 6 basic ones and contempt. Columns: prediction. Rows: ground truth. The average recognition rates are **0.85** for LSTM and 0.80 for SRMP.

6   Conclusions

In this paper, We formulate expression recognition as a video Sparse Representation based Classification (SRC) with Long Short-Term Memory (LSTM) mechanism. The proposed sparse coding with temporal modelling using LSTM successfully detects salient AUs over the time of repeated expressions. With a much lower computation cost due to the selective AUs, our model performs even better (on MPI-VDB) than the baseline of sparse coding with SRMP which analyzes the full visual signal over all frames.

References

1. Xiang, X., Tran, T.D.: Pose-selective max pooling for measuring similarity. In: IAPR International Conference on Pattern Recognition (ICPR) Workshops (2016)
2. Xiang, X., Dao, M., Hager, G.D., Tran, T.D.: Hierarchical sparse and collaborative low-rank representation for emotion recognition. In: IEEE International Conference on Acoustics, Speech and Signal Processing (ICASSP), pp. 3811–3815. IEEE (2015)
3. Chen, S., Donoho, D.: Basis pursuit. In: 1994 Conference Record of the Twenty-Eighth Asilomar Conference on Signals, Systems and Computers, vol. 1, pp. 41–44. IEEE (1994)
4. Cotter, S.F., Rao, B.D., Engan, K., Kreutz-Delgado, K.: Sparse solutions to linear inverse problems with multiple measurement vectors. IEEE Trans. Signal Process. **53**(7), 2477–2488 (2005)
5. Hochreiter, S., Schmidhuber, J.: Long short-term memory. Neural Comput. **9**(8), 1735–1780 (1997)
6. Socher, R., Lin, C.C., Manning, C., Ng, A.Y.: Parsing natural scenes and natural language with recursive neural networks. In: Proceedings of the 28th International Conference on Machine Learning (ICML 2011), pp. 129–136 (2011)

Author Index

Printed in the United States
By Bookmasters